Sound & Vibration

David M. Sykes • Gregory C. Tocci • William J. Cavanaugh
Editors

Sound & Vibration 2.0

Design Guidelines for Health Care Facilities (2010)

Springer

Editors
David M. Sykes
Acoustics Research Council
Boston, Massachusetts
USA

Gregory C. Tocci
Acoustics Research Council
Boston, Massachusetts
USA

William J. Cavanaugh
Acoustics Research Council
Boston, Massachusetts
USA

ISBN 978-1-4614-4986-7   ISBN 978-1-4614-4987-4 (eBook)
DOI 10.1007/978-1-4614-4987-4
Springer New York Heidelberg Dordrecht London

Library of Congress Control Number: 2012946034

© ARC 2013

This work is subject to copyright. All rights are reserved by the Publisher, whether the whole or part of the material is concerned, specifically the rights of translation, reprinting, reuse of illustrations, recitation, broadcasting, reproduction on microfilms or in any other physical way, and transmission or information storage and retrieval, electronic adaptation, computer software, or by similar or dissimilar methodology now known or hereafter developed. Exempted from this legal reservation are brief excerpts in connection with reviews or scholarly analysis or material supplied specifically for the purpose of being entered and executed on a computer system, for exclusive use by the purchaser of the work. Duplication of this publication or parts thereof is permitted only under the provisions of the Copyright Law of the Publisher's location, in its current version, and permission for use must always be obtained from Springer. Permissions for use may be obtained through RightsLink at the Copyright Clearance Center. Violations are liable to prosecution under the respective Copyright Law.

The use of general descriptive names, registered names, trademarks, service marks, etc. in this publication does not imply, even in the absence of a specific statement, that such names are exempt from the relevant protective laws and regulations and therefore free for general use.

While the advice and information in this book are believed to be true and accurate at the date of publication, neither the authors nor the editors nor the publisher can accept any legal responsibility for any errors or omissions that may be made. The publisher makes no warranty, express or implied, with respect to the material contained herein.

Printed on acid-free paper

Springer is part of Springer Science+Business Media (www.springer.com)

SOUND & VIBRATION
Design Guidelines for Health Care Facilities
January 1, 2010

Public Draft 2.0 – including guidelines for NICUs

Prepared for the Facility Guidelines Institute by: **ANSI S12 WG44 & the Joint Subcommittee on Speech Privacy & Healthcare Acoustics TC-AA.NS.SC, a technical committee of the Acoustical Society of America (ASA committees on Architectural Acoustics, Noise & Speech Communication), the Institute of Noise Control Engineers (INCE), & the National Council of Acoustical Consultants (NCAC)**
Co-Chair: David M. Sykes, ASA, INCE
Co-Chair: Gregory C. Tocci, P.E., FASA, F/INCE, INCE Bd. Cert.
Co-Founder: William J. Cavanaugh, FASA, F/INCE, INCE Bd. Cert.

This document is the sole Reference Standard for acoustics by:
- 2010 FGI/ASHE Guidelines for Design and Construction of Health Care Facilities
- Green Guide for Health Care v2.2 (Jan 2007) – 2 acoustics credits
- LEED for Healthcare (in draft) – 2 acoustics credits

NOTE TO READERS:

This document is the sole Reference Standard for architectural acoustics cited by the following organizations:

- FGI/ASHE Guidelines for Design and Construction of Health Care Facilities

    - Green Guide for Healthcare v2.2 (January 2007) which includes 2 new credits for acoustics

    - LEED for Healthcare (in draft) which includes 2 new credits for acoustics

While the technical information is consistently referred to by these organizations, There are variations in how these criteria are used to achieve credits or compliance.

Notes:

1. Error in the 2010 Guidelines:

A new ASTM standard was erroneously referred to in the 2010 Guidelines section 1.1-5.5 Referenced Codes and Standards (ASTM E2638-08). This is an error that will be corrected in an addendum to the Guidelines. This new standard appeared too late in the development process for the 2010 Guidelines to be included in the public review process that preceded publication. Therefore the contents of ASTM E2638-08 were not included in Guidelines *Table 1.5-4: Design Criteria for Speech Privacy.*

2. Corresponding tables in the Guidelines: the FGI/ASHE Guidelines label seven tables from this document as follows:

- **2010 Guidelines Table A1.2-1:** "Categorization of Health Care Facility Sites by Exterior Ambient Sound"

See **page 28** *in this document,* Table 1.3-1: "Categorization of Hospital Sites by Exterior Ambient Sound"

- 2010 Guidelines Table 1.2-1: "Design Room Sound Absorption Coefficients"

See **page 40** *in this document,* Table 2.3-1: "Recommended design room sound absorption coefficients"

- 2010 Guidelines Table 1.2-2: "Minimum-Maximum Design Criteria for Noise in Interior Spaces"

See **page 48** *in this document,* Table 3.3-1: "Recommended criteria for noise in Interior Spaces"

- 2010 Guidelines Table 1.2-3: "Design Criteria for Minimum Sound Isolation Performance Between Enclosed Rooms"

See **page 56** *in this document,* Table 4.3-1: "Recommended Sound Isolation Performance Between Enclosed Rooms"

- 2010 Guidelines table 1.2-4: "Design Criteria for Speech Privacy for Enclosed Rooms and Open Plan Spaces"

See **page 60** *in this document*, Table 4.4-1: "Speech Privacy for Enclosed Rooms" and Table 4-4: "Speech Privacy Goals for Open-Plan Spaces.

- 2010 Guidelines table A2.1-a: "Sound Transmission Loss or Attenuation Through Horizontal and Vertical Barriers in NICUs"

See **page 62** *in this document,* "Adapted from Evans JB, Philbin MK, Facility and operations planning for quiet hospital nurseries"

- 2010 Guidelines table 1.2-5: "Maximum Limits on Footfall Vibration in Health Care Facilities,"

See **page 73** *in this document,* Table 6.3.2-1: "Recommended Limits on Footfall Vibration in Hospitals"

Sound & Vibration Design Guidelines for Health Care Facilities

Contributors

This document was commissioned by and prepared for the Facility Guidelines Institute by ANSI S12 WG44, and the joint subcommittee on speech privacy & healthcare acoustics (TC-AA.NS.SC), a technical committee of the Acoustical Society of America (ASA committees on Architectural Acoustics, Noise & Speech Communication), the Institute of Noise Control Engineers (INCE), and the National Council of Acoustical Consultants (NCAC).

Acoustical Working Group:

William J. Cavanaugh, FASA, F/INCE Bd. Cert.
Gregory C. Tocci, P.E., FASA, F/INCE Bd. Cert.
David M. Sykes, ASA, INCE
Andrew C. Carballeira, ASA
Jeffrey L. Fullerton, INCE
Benjamin C. Davenny, ASA, INCE

Contributing Reviewers:

Douglas H. Bell, ASA, INCE Bd. Cert.
Joost H. Bende, AIA
Susan Blaeser
Timothy G. Brown, ASA
Roger B. Call, AIA
Robert Chanaud, Ph.D.

The Committee to Recommend Standards for Newborn ICU Design
Thomas Corbett
Emily Cross, P.E., ASA
John Erdreich, Ph.D., FASA
Jack Evans, P.E., ASA
Ken Good, ASA
Lewis S. Goodfriend, Ph.D., FASA, INCE Bd. Cert.

Kring Herbert, FASA
Tony Hoover, FASA, INCE Bd. Cert.
Thomas Horrall, FASA
Mandy Kachur, P.E., ASA, INCE Bd. Cert.
Matthew Moore, ASA
Dennis Paoletti, FAIA, FASA
Kathleen Philbin, Ph.D., RN
Kenneth Roy, Ph.D., FASA
Charles Salter, FASA, INCE Bd. Cert.
Nate Sevener, P.E., ASA, INCE Bd. Cert.
Christopher Storch, ASA
Brandon Tinianov, Ph.D., ASA, INCE Bd. Cert.
Eric Ungar, D.Sc., P.E., FASA, F/INCE, INCE Bd. Cert.
Kenric Van Wyck, P.E., ASA, INCE Bd. Cert.

AIA Healthcare Guideline Revision Committee:

Kurt Rockstroh, AIA, ACHA, Vice-Chairman 2010 HGRC
Robert Loranger, P.E., CHFM, 2010 HGRC Steering Committee Member
Dr. Jo Solet, Harvard Medical School, 2010 HGRC Steering Committee Member

Last Revision Date: January 14, 2010

1.6.3 Sound & Vibration

TABLE OF CONTENTS

Note to Readers: 7

Table of Contents 11

Introduction 14

Process Overview & Acknowledgements 17

Development Timeline 19

1 Site Exterior Noise 23

    1.1 General 24

    1.2 Applicable Federal, State, and Local Codes and Regulations 25

    1.3 Classification of non-facility produced exterior noise exposure 26

    1.3.1 Exterior noise classifications 26

    A1.3 Procedure for determining composite sound transmission class rating (STC_c) 29

    1.4 Classification of facility produced exterior noise exposure 32

    1.4.1 Heliports 33

    1.4.2 Emergency power generators 34

    1.4.3 Outdoor mechanical equipment 35

    1.4.4 Building services 35

1.6.3 Sound & Vibration

2 Acoustical Finishes and Details 36

    2.1 General... 37

    2.2 Applicable Federal, State and Local Codes and Regulations 39

    2.3 Design Criteria for Acoustical Finishes 40

    A2.3 Determination of Design Room Average Sound Absorption Coefficient ($\overline{\alpha}_{design}$) and Room Sound Absorption Factors (A_R, sf) 41

    2.4 Considerations for NICUs (from Standard 21: Ceiling Finishes) 43

    A2.4.1 Interpretation 44

    A2.4.2 Interpretation: 44

3 Room Noise Levels 45

    3.1 General... 46

    3.2 Federal, State and Local Codes, Regulations, and Guidelines 46

    3.3 Design Criteria for Room Noise Levels 47

    3.4 Conformance measurements of room sound level 49

    A3.4.1 Determination of Room Noise Level 49

    A3.4.2 Effect of Background Noise on Clinical Hearing Ability 49

    A3.4.3 Discussion of Background Noise Rating Criteria 49

    3.5 Considerations for NICUs (from Standard 23: Acoustic Environment) 52

    A3.5.1 Interpretation: 52

    A3.5.2 Interpretation: 52

4 Sound Isolation Performance of Constructions 54

    4.1 General... 55

    4.2 Applicable Federal, State and Local Codes and Regulation............................ 55

    4.3 Design guidelines for sound isolation between enclosed rooms 55

    A4.3.1 Typical partitions 57

    A4.3.2 Extraordinary partitions 57

    A4.3.3 Composite Sound Transmission Class (STCc)... 59

    4.4 Design guidelines for speech privacy between enclosed rooms 60

    A4.4 Enclosed Room Speech Privacy Design Guidance 61

    4.5 Design guidelines for speech privacy in open-plan spaces 61

    4.6 Open Plan Speech Privacy Design Guidance 61

1.6.3 Sound & Vibration

    4.6.1 Considerations for NICUs (from Standard 23: Acoustic Environment)..........62

    4.6.2 Considerations for NICUs (from Standard 23: Acoustic Environment)..........62

    Speech Privacy References: 63

    A4.6 Interpretation: 63

5 Paging & Call Systems, Clinical Alarms, Masking Systems & Sound Reinforcement 65

    5.1 General...66

    5.2 Applicable Federal, State and Local Codes and Regulation..............................66

    5.3 Paging and Call Systems 66

    5.4 Clinical alarms 67

    A5.4 Audibility of tonal alarms 68

    5.5 Masking Systems 68

    5.6 Sound Reinforcement...68

    5.7 Considerations for NICUs (from Standard 23: Acoustic Environment) 69

    A5.7 Interpretation: 69

    References..69

6 Building Vibration 70

    6.1 General...71

    6.2 Applicable Federal, State and Local Codes and Design Guides.......................72

    6.3 Vibration Control and Isolation 72

    6.3.1 Mechanical, Electrical and Plumbing Equipment (MEP) 72

    6.3.2 Structural 72

    6.3.3 Structureborne sound 73

    6.3.4 Medical and laboratory instrumentation 73

Glossary 74

Partial List of Abbreviations 83

References 85

    New references: 85

INTRODUCTION

1.6.3 Sound & Vibration

INTRODUCTION

The recommended minimum design requirements in this document are intended to assure *satisfactory acoustical and privacy environments* in healthcare facilities of all types, including large general hospitals, specialized patient care facilities, and ambulatory patient care facilities, as well as other healthcare venues specified in the *2010 FGI/ASHE Guidelines for the Design and Construction of Healthcare Facilities*.

A satisfactory acoustical environment in a healthcare facility is one in which all the sounds are compatible with the intended use of the spaces, within the framework of the *Environment of Care* concept described by the Facility Guidelines Institute.

Sensible acoustical & privacy planning in the early design stages of a healthcare facility project can be solved effectively and affordably with "a few strokes of the designer's pencil." These *Sound and Vibration Design Guidelines* are therefore intended to aid designers in achieving *satisfactory acoustical and privacy environments* in all types of healthcare facilities, whether new or renovated.

This document was prepared by the Joint Subcommittee TC-AA.NS.SC and ANSI S12 WG44 (identified jointly herein as "The Acoustical Working Group"). The committees were formally organized in 2004-5 with members drawn from ten constituencies ranging from medicine to law and public policy to engineering and design in order to provide constructive guidance on sound and vibration to the Facilities Guidelines Institute (FGI) and provide a Reference Standard for other groups such as the Green Guide for Healthcare and LEED.

In August 2005, FGI commissioned the development of this document stipulating that it should be: (1) *comprehensive*, (2) *practical* in nature, (3) based on *existing technical standards* from recognized authorities, (4) based on generally-accepted *professional best practices* in the acoustics industry, (5) achievable using *available products and methods*, and (6) advised by *recent, reputable, evidence-based and/or clinical research*.

1.6.3 Sound & Vibration

The goal was to prepare the first set of comprehensive, practical, measureable guidelines for all aspects of acoustics in the design, construction and evaluation of all types of healthcare facilities whether new, renovated or rehabilitated.

The overall mission of the Facility Guidelines Institute is to "develop performance-oriented *minimum requirements* as suggested standards for American health care facility design without prescribing design solutions" (2006 guidelines, page 3).

Sound and vibration are topics of increasing prominence in the design, construction, and operation of health care facilities. But the subject is not new. Several recent articles provide summaries of the extensive literature available on acoustics in health care facilities. Three articles worth noting for their extensive bibliographies are: "The role of the physical environment in the hospital of the 21st century: a once-in-a-lifetime opportunity" by Ulrich et al. (Center for Health Design, 2004); "Noise levels in Johns Hopkins Hospital" by Busch-Vishniac et al. (JASA, 2005); and "Sound Control for Improved Outcomes in Healthcare Settings" by Joseph & Ulrich (Center for Health Design, 2007).

Sound and "acoustical comfort" in buildings can be practically achieved and measurably controlled in a number of ways that are proven: with room finish materials that absorb sound energy, with exterior and interior walls and floor/ceiling structures that block sound transmission, with vibration-isolating equipment supports, and a vast arsenal of other methods known to engineers and designers.

One common sense approach is simply to specify quiet equipment—this is oftentimes the least costly, most effective strategy for controlling noise or vibration. Another common-sense approach consists of adding a low level of continuous background sound--a "positive distraction"--that can actually be the most effective way of achieving speech privacy and "acoustical comfort."

Further, the judicious placement of acoustical "buffer spaces" (corridors, storage, and other non-critical rooms) around critical or sensitive medical testing spaces may save considerably on partitioning costs for high performance sound isolation wall constructions.

Whatever the specifics, there are proven, objective, scientific and engineering methods available for *measuring, mitigating, monitoring and controlling* noise in healthcare facilities and these methods are well established, having been practiced worldwide for over a century by thousands of engineers, architects and designers.

The Committee Chairs

January 2010

1.6.3 Sound & Vibration

PROCESS OVERVIEW & ACKNOWLEDGEMENTS

This is Public Draft 2.0 (January 2010) of *Sound and Vibration: Design Guidelines for Hospitals and Healthcare Facilities*. It supersedes the previous drafts, Public Draft 1.0 (November 2006) and Public Draft 1.1 (2008 – incorporating NICU requirements). Public Draft 2.0 continues to be a work in progress on which comments are actively solicited from a wide variety of constituencies for the purpose of revising the 2016 Guidelines, which are already a work in progress.

Drafting of this document was formally requested by the Board of Directors and Steering Committee of the Facilities Guidelines Institute on September 2, 2005. The first complete draft (5a) was promised and delivered within nine months by mid-June 2006, in time for distribution at the meeting of the Institute Steering Committee in Washington, DC. A revised version of Public Draft 1.0 was delivered to the Institute on November 6, 2006 and underwent several rounds of public review.

This document was prepared by a team of volunteer contributors representing the 500-member, international *ASA Joint Technical Subcommittee on Speech Privacy*, also known as *ANSI S12/WG44 – Healthcare Acoustics & Speech Privacy*. The committee's members represent nine constituencies including: federal agencies in the USA and Canada, academic institutions, healthcare institutions, several professions including medicine, law, healthcare policy & management, architecture & planning, engineering, acoustical science, engineering & consulting, facilities management, and materials/equipment manufacturing.

The Co-chairmen of the Joint Subcommittee, Gregory Tocci and David Sykes, and its co-founder, William Cavanaugh, actively participated throughout the drafting process and guided each of the monthly progress meetings that were held in Boston MA in the presence of designated representatives of the Steering Committed of the Facility Guidelines Institute.

1.6.3 Sound & Vibration

The team included representatives of a number of professional associations in healthcare policy (WEDI SNIP), healthcare law (AHLA), healthcare architecture (AAH), pediatrics (AAP), healthcare engineering (ASHE), and the three leading professional engineering organizations in acoustics: the ASA (Acoustical Society of America) - Technical Committees on Architectural Acoustics and Noise; INCE (Institute of Noise Control Engineering); and NCAC (National Council of Acoustical Consultants).

Special thanks

The drafting group was privileged to work under the guidance of two members of the Steering Committee of the Facility Guidelines Institute, Kurt Rockstroh, Co-chair of the 2010 Guidelines, and Robert Loranger, a member of the Institute Steering Committee. In the nearly forth monthly meetings held in Boston to review progress drafts between September 2005 and January 2010, their advice and counsel were always constructive and provided invaluable insight and experience based on their past experience with developing previous editions of the *Design Guidelines for Healthcare Facilities*.

The committee greatly appreciates the assistance of Doug Erickson, president of the Facility Guidelines Institute who welcomed and encouraged the development of these guidelines, and Martin Cohen, Co-chairman of the *2010 Guidelines*, who attended a special convocation organized by the committee during the 151st meeting of the Acoustical Society of America on June 7, 2006 in Providence, Rhode Island where he gave a presentation about the *AIA Healthcare Guidelines* and the work of the Facility Guidelines Institute.

We also greatly appreciate the interest, encouragement and assistance of Sholem Prasow, a member of the Core Committees of GGHC and LEED HC who welcomed, encouraged and shepherded the adoption of these guidelines by these organizations as their sole "reference standard" for acoustics. Prasow was aided in this by four new members of the USGBC's Environmental Quality Technical Advisory Group (EQ-TAG) who are themselves professionals in acoustics: Dan Bruck, Alexis Kurtz, David Lubman and Charles Salter.

Credits

Design: Marc Drucker, Newlogic, Boston MA

Photographs: all photographs supplied by Steffian Bradley Architects, Boston MA

1.6.3 Sound & Vibration

DEVELOPMENT TIMELINE

This chronology was prepared at FGI's request to provide an overview of the process by which the acoustical "Interim guidelines" were developed.

2002 – Committee co-chair (Sykes) gives testimony to two congressional committees about the HIPAA requirement to protect "oral communications" for which no guidance had yet been prepared. Matter referred to the statutory advisor for HIPAA, WEDI-SNIP.

2003 - April 2004 – Committee co-chair (Sykes) and WEDI-SNIP co-chair (Susan Miller PhD) draft the first white paper on "Oral Commucations/Speech Privacy" for the HIPAA statutory advisory group WEDI-SNIP (Workgroup on Electronic Data Interchange), published in April 2004, coinciding with the release of the HIPAA Privacy Rule.

December 2004 – ASA approves co-chair Tocci's request to form the Joint Technical Subcommittee on Speech Privacy. The Co-chairmen issue a call for participation in drafting an Interim Sound and Vibration Design Guideline for Hospital and Healthcare Facilities and begin building constituencies and professional outreach.

December 2004 – First contact between the committee co-chair (Sykes) and FGI (Erickson) to discuss development of acoustical guidelines.

May 2005 – First official meeting of the Joint Subcommittee at the 149th meeting of ASA in Vancouver BC, Canada. Co-chairs (Tocci, Sykes, Cavanaugh) request a special session at the 150th meeting of the ASA in Minneapolis, Minnesota in October and issue a call for papers on speech privacy and healthcare acoustics.

1.6.3 Sound & Vibration

August 2005 – Subcommittee co-chair (Sykes) attends the Steering Committee meeting of the Facilities Guidelines Institute (FGI) in Washington DC to propose the drafting of an *Interim Sound and Vibration Design Guideline for Hospital and Healthcare Facilities*. FGI Steering Committee calls for board members to form an oversight committee (Rockstroh & Loranger) to supervise this process.

September 2005 – The FGI Steering Committee approves drafting of an *Interim Sound and Vibration Design Guideline for Hospital and Healthcare Facilities* to be prepared by the ASA Joint Technical Subcommittee with a tentative delivery date of June 2006.

October 2005 – Second meeting of the Subcommittee at the ASA meeting in Minneapolis; tasks outlined, call for volunteers to be involved in researching and drafting the proposed *Interim Sound and Vibration Design Guideline for Hospital and Healthcare Facilities*. Subcommittee organizes website (www.healthcareacoustics.org & www.speechprivacy.org) as vehicle for communication, outreach, and constituency building, and petitions ANSI for designation as S12 Workgroup 44.

November 2005 – First drafting meeting held in Boston MA to outline the purpose, scope and schedule of the proposed document.

January to May 2006 – monthly scheduled meetings held in Boston between the drafting group and the FGI oversight committee to review progress drafts.

March 2006 – ASA Joint Technical Subcommittee designated ANSI S12 Work Group 44 – Healthcare Acoustics & Speech Privacy.

April 2006 – Research team begins preparing and submitting grant proposals to conduct transdisciplinary clinical validation studies under the auspices of Harvard Medical School.

June 2006 – the ASA Joint Technical Subcommittee (ANSI S12 Work Group 44) assembles a special convocation of fourteen subject matter pioneers, leaders and experts at the 151st ASA meeting in Providence, RI called "Speech Privacy: a 50th Anniversary Celebration" and also holds its third official meeting, proposing formation of a clinical research program to validate the recommendations in the *Interim Sound and Vibration Design Guideline for Hospital and Healthcare Facilities*. FGI co-chair Cohen presents.

1.6.3 Sound & Vibration

June 2006 - The drafting group delivers Draft 5a to the FGI Steering Committee for distribution at the Steering Committee meeting in Washington, DC. Simultaneously, the ASA Joint Technical Subcommittee issues a call for peer review and comment by its members.

July 2006 – FGI leaders (Erickson, Rockstroh) meet with drafting team and researchers from Harvard Medical School to discuss next steps and research.

September 15, 2006 – public awareness effort begins with an article about the *Interim Sound and Vibration Design Guideline…* in *Healthcare Design* magazine (www.hcdmagazine.com), co-authored by Sykes, Rockstroh, Solet and Buxton.

September 30, 2006 – comment period closed for members of the ASA Joint Technical Subcommittee; drafting group begins weekly meetings to compile and review comments.

October 24, 2006 – Revised draft 5c presented to the FGI oversight committee for discussion and comment.

October, 2006 - Research team organized by the committee receives notice of its first grant award for clinical validation research.

November 2006 – Co-chairman (Sykes) meets with board member of the USGBC's - LEED for Healthcare Committee/Healthcare EQ Acoustics Credit subcommittee in Toronto, CN to discuss adoption by GGHC and LEED of the FGI's newly completed acoustical criteria.

November 4, 2006 – Revised "Public Draft 1" submitted to FGI Steering Committee for co-distribution by the ASA Joint Technical Subcommittee/ANSI S12 Work Group 44 and by FGI to the members of its 120-member national Health Guidelines Revision Committee. Healthcare industry peer review officially begins. FGI forms subcommittee to review & format the acoustical guidelines in a manner consistent with the recently completed *2006 AIA Guidelines for Hospital and Healthcare Facilities.*

November 5, 2006 – LEED HC core committee member requests committee's approval to adopt the 2006 Interim Guidelines as the "Reference Standard" for acoustics and privacy in GGHC v2.2 (pilot phase for LEED HC), and LEED HC.

1.6.3 Sound & Vibration

December 2006 – peer review continues among engineering professions as the ASA Joint Subcommittee holds its fourth international meeting in Honolulu, HI at the 152nd ASA meeting. Subcommittee co-chairmen deliver an invited paper at the conjointly-held INCE -InterNoise conference with co-authors Rockstroh & Solet.

January 31, 2007 – The Green Guide for Healthcare (GGHC) adopts the Interim Guideline as the sole Reference Standard for two Environmental Quality credits in its version of LEED HC.

December 2006 to February 2007 – comments from parallel peer review processes funneled to drafting team in preparation for next draft.

February 2007 – Center for Health Design publishes its white paper on healthcare acoustics referencing the Interim Guideline, "Sound Control for Improved Outcomes in Healthcare Settings" (A. Joseph PhD and R. Ulrich PhD).

February 2007 – Interim Sound and Vibration Design Guideline for Hospital and Healthcare Facilities discussed at ASHE PDC (division of the American Hospital Association) in San Antonio, TX by a panel consisting of Sykes, Cavanaugh, Rockstroh & Solet.

April 2007 – Interim Sound and Vibration Design Guideline for Hospital and Healthcare Facilities "Public Draft 2" presented to the first full meeting of the 2010 Healthcare Guidelines Revision Committee in Baltimore/Washington area.

November 2007 - Committee co-chairs present the acoustical criteria at the U.S.Green Building Council's "Greenbuild" conference in Chicago, IL.

2008 – Committee co-chairs present the acoustical criteria at several conferences including the HIPAA Privacy Conference at Harvard University and the AIA National Conference in Boston MA. The committee also organizes a special symposium in Paris, France focused on acoustics and privacy in healthcare facilities.

2009 – Several rounds of peer review and public review of the document followed by corrections and changes. Committee co-chairs present the acoustical criteria at several national conferences including the Center for Health Design's national conference in Orlando, FL.

January 2010 – FGI publishes the 2010 Guidelines; simultaneously the acoustics committee publishes "Sound & Vibration Design Guidelines for Health Care Facilities" dated January 1, 2010 as the reference standard for the 2010 Guidelines.

1 SITE EXTERIOR NOISE

1.6.3.1 Sound & Vibration - Site Exterior Noise

1 SITE EXTERIOR NOISE

1.1 General

(1) This section provides design guidance on how to address environmental noise at a facility site over which the facility may or may not have administrative or operational control. This section is intended to serve as a means for preliminary assessment of the suitability of a site with respect to environmental noise exposure.

   (a) Examples of noise sources that a facility can control are power plant, HVAC equipment, and emergency generators that are part of the health care facility.

   (b) Examples of noise sources that a facility cannot control include highways, rail lines, and airports, and general urban, industrial, and public service equipment and activities.

   (c) There are other sources over which the facility may have limited control such as helipads. Guidelines for location and operation of helipads are subject to federal regulation and other safety and environmental considerations.

(2) The planning and design of new facilities or the retrofitting of existing facilities shall include due consideration of all existing exterior noise sources that may be transmitted from outside a building to its interior through the exterior shell (exterior walls, windows, doors, roofs, ventilation and other shell penetrations). Hospital design should also consider future noise source development, such as the construction of highways, airports, or rail lines in the vicinity of the project.

(3) Planning and design shall also include due consideration of hospital facility noise source sound transmission to nearby residences and other sensitive receptors. Exterior facility equipment sound can be controlled to achieve acceptable sound levels inside hospital facility spaces and at neighboring receptors through compatible siting of noise sources and receptors to take

1.6.3.1 Sound & Vibration - Site Exterior Noise

advantage of distance, orientation, and shielding. Exterior facility equipment sound can also be controlled using quiet equipment selections and noise control equipment such as silencers and barriers.

1.2 Applicable Federal, State, and Local Codes and Regulations

The planning and design of new facilities or the retrofitting of existing facilities shall conform to all applicable codes, regulations, and guidelines setting limits on exterior sound in the environment and interior sound within all building spaces.[1]

Applicable codes, regulations, and guidelines include:

(1) Department of Health and Human Services (including HIPAA)
(2) AAP Guidelines for Noise in NICUs
(3) Federal Aviation Administration (FAA, for helipad design, construction, and operation)
(4) Building code of the local or state jurisdiction.
(5) Local and state limits on environmental sound.
(6) Local planning and zoning limits on environmental sound.
(7) Occupational Safety and Health Administration—OSHA (worker noise exposure in areas where sound levels exceed 85 dBA)
(8) Professional society design guidelines for noise (American Society of Heating, Refrigerating, and Air-Conditioning Engineers - ASHRAE for mechanical system sound and vibration control)
(9) American National Standards Institute—ANSI (guideline for sound in building spaces and special spaces such as booths for measuring hearing threshold)
(10) Manufacturer guidelines for sound and vibration sensitive medical equipment or equipment that produces sound and/or vibration.

---

[1] When codes set limits they usually are expressed as maximum A-weighted sound levels in dBA. Often separate limits are set for day and night periods, where the nighttime limit is typically 5 to 10 dBA lower than the daytime limit. Daytime limit typically vary between 55 and 65 dBA.

1.6.3.1 Sound & Vibration - Site Exterior Noise

1.3 Classification of non-facility produced exterior noise exposure

1.3.1 Exterior noise classifications

(1) By means of exterior site observations and/or a sound level monitoring survey, the facility site shall be classified into one of the noise exposure categories of Table 1.3-1. The sound levels provided in Table 1.3-1 for noise exposure categories A through D are intended to be used for the evaluation of required healthcare building envelope sound isolation and may differ from other such categorizations of community noise made elsewhere in this document.

    (a) Category A – Minimal environmental sound. As typified by a rural or quiet suburban neighborhood with ambient sound suitable for single family residences, and where sound produced by transportation (highways, aircraft, and trains) or industrial activity may be occasionally audible, and are only minor features of the acoustical environment.

    (b) Category B – Moderate environmental sound. As typified by a busy suburban neighborhood with ambient sound suitable for multifamily residences, where sound produced by transportation or industrial activity are clearly audible and may at time dominate the environment, but are not so loud as to interfere with normal conversation outdoors.

    (c) Category C – Significant environmental sound. As typified by a commercial urban location, possibly with some large apartment buildings, where sound produced by transportation or industrial activity dominates the environment and often interferes with normal conversation outdoors.

    (d) Category D – Extreme environmental sound. As typified by a commercial urban location immediately adjacent to transportation or industrial activities, where sound nearly always interferes with normal conversation outdoors.

(2) Environmental noise on Category B, C, and D sites shall be evaluated generally using the methods of ANSI S12.9 Part 2 for continuous sound monitoring over a minimum 1-week period to document site ambient sound levels. This information is needed to determine detailed environmental noise control requirements for building design. Sites having ambient sound influenced by airport operations may require additional monitoring as suggested in the ANSI standard to account for weather related variations in aircraft sound exposure on-site. In lieu of this additional monitoring, aircraft sound level contours may be available from the airport that can be used to determine the day-night average sound level on-site produced by aircraft operations associated with the nearby airport. Sound level monitoring on-site would still be needed to determine sound levels produced by other sources as well.

1.6.3.1 Sound & Vibration - Site Exterior Noise

(3) For exterior sound exposure categories A through D, Table 1.3-1 presents general descriptions including distance from major transportation noise sources, ambient sound levels produced by non-hospital sound sources, and corresponding design goals for the sound isolation performance of the exterior building shell.

(4) The exterior shell composite STC ratings in Table 1.3-1 are for windows closed. Opening windows effective reduces shell composite STC ratings to 10 to 15, depending on the amount windows are opened. Consideration must be given to whether windows would be opened, and for how long and under what circumstances and the potential impact must be identified in the design.

(5) The outdoor sound levels, expressed as A-weighted day-night average sound levels, are provided in the context of exterior building shell design. Outdoor patient areas would like require lower sound levels typically not exceeding a day-night average sound level of 50 dB. This may require exterior noise barriers or locating outdoor patient areas to gain shielding of noise sources by building structures for example.

1.6.3.1 Sound & Vibration - Site Exterior Noise

Guidelines Table: 1.2-1: Categorization of Health Care Facility Sites by Exterior Ambient Sound

Table 1.3-1: Categorization of hospital sites by exterior ambient sound

Exterior Site Noise Exposure Category	A	B	C	D
General description	Minimal	Moderate	Significant	Extreme
Day-night average sound level (L_{dn}) (dB)	< 65	65-70	70-75	> 75
Average hourly nominal maximum sound level (L_{01}) (dBA)	< 75	75-80	80-85	> 85
Distance from nearest highway (ft)	1000	250-1000	60-250	< 60
Slant distance from nearest aircraft flight track [1] (ft)	> 7000	3500-7000	1800-3500	< 1800
Distance from nearest rail line (ft)	1500	500-1500	100-500	< 100
Exterior shell composite STC rating[2] (STC_c)	35	40	45	50
Exterior patient sitting areas	Some shielding of principal noise sources may be required	Requires shielding of principal noise sources	Generally not acceptable without special acoustical consideration	Generally not acceptable

Notes:

[1] The shortest distance occurring in an over flight between an aircraft and a hospital building. Slant distances have been selected to result in L_{dn}s that are 10 dB below the category range to account for the disproportionate sensitivity of communities to aircraft noise as compared to other sources of environmental noise such as traffic.

[2] Outdoor-Indoor Transmission Class (OITC) rating as defined in ASTM E 1332 is specially suited for evaluating the sound isolation performance of exterior building facades and components. However, currently, this information is not often provided manufacturers. The STC rating is the more widely used method for evaluating sound isolation of all building components.

1.6.3.1 Sound & Vibration - Site Exterior Noise

A1.3 Procedure for determining composite sound transmission class rating (STC_c)

In the previous section, a method for determining the minimum required shell composite STC rating has been presented. This section discusses how the STC rating for the exterior wall and the size of windows to be installed can be used to determine the minimum required window STC rating such that the composite STC rating of the wall/window assembly as a whole will achieve the minimum required STC rating determined from the previous section.

This method for determining wall/window composite STC has been arranged to recognize that the building wall system is usually selected first and the window assembly follows. Note that the window is usually the most significant sound transmitting component of a building shell.

(1) From Table 1.3-1 determine the site exposure category and the corresponding composite STC rating (STC_c).
(2) Determine the percentage of the exterior wall occupied by the window or windows.
(3) Select a window assembly. Determine its STC rating from manufacturer or other reference.
(4) Use Table 1.3-2 to determine minimum required wall STC rating.
(5) If minimum required window STC rating is shown as "n/a" then choose a wall with a higher STC rating. Continue increasing wall STC rating until a window STC rating is shown in the appropriate table cell. This may also require reducing the window area as well.

1.6.3.1 Sound & Vibration - Site Exterior Noise

Table A1.3-2: Minimum required wall STC rating for combinations of required composite shell STC rating, window STC rating, and window area

STC_c	STC_{wall}	Window Area as a Percentage of Total Wall Area			
		10%	20%	40%	80%
		STC_{window}			
35	40	26	29	32	34
	45	25	28	31	34
	50	25	28	31	34
	55	25	28	31	34

STC_c	STC_{wall}	Window Area as a Percentage of Total Wall Area			
		10%	20%	40%	80%
		STC_{window}			
40	40	40	40	40	40
	45	31	34	37	39
	50	30	33	36	39
	55	30	33	36	39

STC_c	STC_{wall}	Window Area as a Percentage of Total Wall Area			
		10%	20%	40%	80%
		STC_{window}			
45	40	n/a	n/a	n/a	48
	45	45	45	45	45
	50	36	39	42	44
	55	35	38	41	44

STC_c	STC_{wall}	Window Area as a Percentage of Total Wall Area			
		10%	20%	40%	80%
		STC_{window}			
50	40	n/a	n/a	n/a	n/a
	45	n/a	n/a	n/a	53
	50	50	50	50	50
	55	41	44	47	49

1.6.3.1 Sound & Vibration - Site Exterior Noise

Table A1.3-3: Typical wall STC ratings.

Typical wall STC ratings:	
EIFS	40-55
Lightweight exterior finish, heavy gauge metal stud with glass fiber in cavity, resilient channels on inner side of stud frame and one layer of GWB	45
Same as above except with resilient channels on the inside of wall frame	50
Heavy masonry exterior, moisture control air space with metal ties, heavy gauge metal stud frame, glass fiber batt, separate or staggered interior metal stud frame, 2 layers GWB on the interior	60

Table A1.3-4: Typical window STC ratings

Typical window STC ratings[1]:	
1" standard insulating glass (1/4" glass – ½" air space – ¼" glass)	35
1" laminated insulating glass (1/4" lam. glass – ½" air space – ¼" glass)	39
1" double laminated insulating glass (1/4" lam. glass – ½" air space – ¼" lam. glass)	42
4" insulating glass (1/4" glass – 3.5" air space – ¼" glass)	45

Notes:

[1] These STC ratings assume either fixed windows or well sealing operable windows. Operable windows glazed with the configurations indicated, that do not seal well, may not achieve the STC ratings shown.

STC ratings for additional wall and window assemblies available from the following:

- Catalog of STC and IIC ratings for Wall and Floor/Ceiling Assemblies, State of California Department of Health Services, 1981
- Harris, Cyril M., Noise Control in Buildings, McGraw-Hill Book Company, NY 1994.
- Gypsum Board Walls: Transmission Loss Data, National Research Council of Canada Institute for Research in Construction publication 761, March 1968.
- Saflex Acoustical Design Guide, 1999 at http://www.saflex.com/pages/technical/acoustical.asp

1.6.3.1 Sound & Vibration - Site Exterior Noise

1.4 Classification of facility produced exterior noise exposure

Table 1.4-1. Design goals for hospital equipment sound at nearest residences according to exterior site noise exposure category

Exterior Site Noise Exposure Category	A	B	C	D
General description	Minimal	Moderate	Significant	Extreme
Exterior envelope composite STC rating	≥ 35	≥ 40	≥ 45	≥ 50
Exterior patient sitting areas	Some shielding of principal noise sources may be required	Requires shielding of principal noise sources	Generally not acceptable without special acoustical consideration	Generally not acceptable
Design goal for hospital nighttime exterior equipment sound[1] (dBA)	45	50	55	60

Notes

[1] Transmitted to adjacent residential receptors, in the absence of a local code. For equipment operating during daytime only, levels may be increased by 5 dBA. These values differ from sound levels in Table 1.3-1 which are used in the evaluation of required building envelope sound isolation.

1.6.3.1 Sound & Vibration - Site Exterior Noise

1.4.1 Heliports

The location of heliports on a hospital site shall be evaluated for noise impacts on the facility and community. Helipads can be located at ground level on the hospital site or on a roof of a hospital building. Helicopter noise at nearest residences and at hospital buildings requires special consideration under the following conditions:

(1) When helicopter sound levels exceed 80 dBA at nearby residences.[2]

(2) When the number of helicopter operations exceeds 3 per day.

(3) When there is likely to be more than 2 helicopter flights per week between the hours of 10:00 P.M. and 7:00 A.M.

(4) When the slant distance to the nearest residence is 1000 feet or less.

(5) When the helipad is atop a hospital building, special attention to the design of windows of the building is required.[3]

(6) When the helipad is located on the ground and situated such that helicopters approach within 500 feet of hospital buildings.

(7) When military helicopters, which are often larger than civilian medi-vac helicopters, are expected to use the helipad more than once per week.

(8) Helicopters, particularly military helicopters and large civilian helicopters, can induce low frequency vibration

(9) in building windows and facades that can produce rattling of building fixtures and furnishings. On the one hand, such rattling is generally not acceptable, but on the other it can be difficult to predict. As a guide, rattling can occur when low frequency sound levels (16-31 Hz) exceed 75 dB and when helicopters are within 500 feet of buildings.

---

[2] This generally occurs when the slant distance from the helicopter to the residence is 700 feet or less. Slant distance is the minimum distance in feet directly between a residence and a helicopter at its closest approach from the residence. Patient transport agencies expecting to use the heliport can provide guidance on slant distances for various helicopter approaches. Helicopter approaches to a helipad are influenced by wind direction and locations of nearby buildings.

[3] Sound levels at windows directly below the flight path to the roof can exceed 90 dBA and may require special acoustical glazing.

1.6.3.1 Sound & Vibration - Site Exterior Noise

1.4.2 Emergency power generators

(1) When an emergency generator is located outdoors, it shall be within a sound reduction enclosure as follows:[4,5]

    (a) Select an enclosure that reduces sound at hospital building facades to a level not exceeding 70 dBA, and to a sound level not exceeding the applicable community noise code at nearest residences for the period of day when maintenance operations occur.[6]

    (b) Locate engine exhaust muffler within the emergency generator enclosure.

(2) When an emergency generator is located within a hospital building, interior noise goals of Section 3 Room Noise Levels and exterior noise goals of Section 1 Site Exterior Noise Exposure Classifications shall be met.[7]

---

[4] Health care facilities generally have emergency power sources for life-safety systems. These currently are most often diesel engine driven electrical generators located either outside or inside a building. Some systems may be provided with battery backup, permitting equipment and systems operations only long enough for safe shut-down. As medical facilities become more sophisticated, the number of systems provided with emergency power in new facilities is increasing, and emergency power provided in existing hospital facilities is expanding, as well.

[5] Manufacturers have come to rate their enclosures as providing 20 to 40 dBA of noise reduction (usually in 5-dBA steps), i.e. a further reduction above that provided by a weatherproof enclosure alone. Emergency generator enclosures identified as for weatherproof purposes only typically provide a 5 to 10 dBA reduction at most.

[6] This often translates into an emergency generator enclosure rated to provide a 30 to 35 dBA noise reduction.

[7] If over occupied space, the emergency generator room will require a concrete floated floor supported on elastomeric isolators.

Support the engine generator atop the floated floor on restrained (un-housed) steel spring vibration isolators sized to provide a minimum 2 inch static deflection.

The engine exhaust muffler shall provide a minimum attenuation of 40 dB at 125 Hz.

Install duct silencers on the room air intake to and discharge from the emergency generator room.

Spaces surrounding the emergency generator room shall be utility or mechanical spaces unless special sound isolation constructions are employed.

1.6.3.1 Sound & Vibration - Site Exterior Noise

1.4.3 Outdoor mechanical equipment

Outdoor mechanical equipment includes cooling towers, rooftop air handlers, exhaust fans, and fans located inside buildings with openings on the outside of the building. Noise that these and other outdoor equipment produce may impinge on hospital buildings and may require special consideration of the hospital building shell in these areas, or may impinge on adjacent properties where jurisdictional noise limits and/or owner land uses must be considered.

Outdoor mechanical equipment shall not produce sound that exceeds:

(1) 65 dBA at the hospital façade unless special consideration is given to façade sound isolation design in impinged areas.
(2) Local daytime and nighttime noise limits at neighboring properties.
(3) Noise goals of 1.4 Site Exterior Noise Exposure Classifications in the absence of a community noise limit.

1.4.4 Building services

Building services include trash compacting and removal, truck unloading, refrigeration trucks, and ambulance arrival. These all include potentially noisy diesel vehicles entering into, under, or immediately adjacent to the building. Hours of operation are usually unlimited. Patient or other sensitive rooms often overlook these areas.

The transmission of sound that these vehicles and associated activities produce into the building shall be considered in a fashion consistent with the goals and guidelines of this document.[8]

---

[8] It is common for ambulances to shut-off sirens before arrival on hospital environs. However, certain ambulance services may be required to use sirens up to the ambulance arrival area.

Sound produced by building service vehicles and equipment may require the installation of windows with enhanced sound isolation in sensitive rooms overlooking these areas. Enhanced sound isolation may take the form of acoustical windows or interior acoustical glazing. Sound isolation windows usually have STC ratings exceeding 40. Interior acoustical sash is usually an interior light of ¼" laminated glass glazed into a separate frame inside the prime window such that a 3 inch or greater clearance is achieved between the prime glass and the interior acoustical sash glass. Insulating glass is usually two lights of annealed glass bonded together using a poly vinyl butyrl interlayer such as Saflex by Solutia.

2 ACOUSTICAL FINISHES AND DETAILS

1.6.3.2 Sound & Vibration - Acoustical Finishes and Details

2 ACOUSTICAL FINISHES AND DETAILS

2.1 General

(1) Room finishes which absorb sound throughout the audible frequency range shall be considered for all occupied spaces throughout the facility. For example,

    (a) surface applied ceiling or wall panels

    (b) suspended acoustical ceiling systems

    (c) floor carpeting

    (d) drapery, furniture, etc.

(2) For these design guidelines, room finishes shall be classified **acoustical** when the mid-frequency sound absorption coefficient (α_{500}) or the frequency averaged noise reduction coefficient (NRC) is not less than 0.5.[9,10]

(3) Typical mid-frequency sound absorption coefficients (α_{500}) and their corresponding NRC values are given in tables 2-1 and 2-2

---

[9] The sound absorption coefficient, α, is the fraction of incident sound energy absorbed (i.e. not reflected back into the source room). α is a dimensionless value ranging from 0 (no absorption, or perfect reflection) to 1 (100% absorption).

[10] NRC is the arithmetic average of the measured sound absorption coefficients at 250, 500, 1000, and 2000 Hz rounded to the nearest 0.05.

1.6.3.2 Sound & Vibration - Acoustical Finishes and Details

Table 2-1: Typical absorption coefficients for common *non-acoustical* building finishes.

Material	α 500	NRC
Brick, unglazed	0.03	0.05
1/2" GWB on metal studs	0.05	0.05
Carpet on concrete	0.14	0.30
Concrete floor	0.01	0.00
Wood floor	0.11	0.10
Glass, ¼" plate	0.04	0.05
½" plywood panels	0.17	0.15
Thin drapery fabric	0.11	0.15
See acoustical texts or handbooks for additional building materials		

Table 2-2: Typical sound absorption coefficients of *acoustical* finishes.[11]

Acoustical Material	α 500	NRC
3/4" fissured mineral tile	0.53	0.70
Suspended glass fiber ceiling	0.70	0.90
Carpet on pad	0.57	0.55
Heavy velour drapes	0.49	0.55
2" wood fiber acoustical deck	0.62	0.60
8" slotted acoustical masonry	0.86	0.65
See manufacturer's literature for additional acoustical products.		

---

[11] Measured sound absorption coefficients for common building and acoustical finishes are typically available in acoustical textbooks and in manufacturer's literature.

1.6.3.2 Sound & Vibration - Acoustical Finishes and Details

(4) In general, acoustical finishes on the boundary surfaces of enclosed rooms and in open plan spaces provide significant reduction of the persistence of and perceived "noisiness" of sounds generated in or transmitted to those spaces such as:

   (a) Conversational speech, loud talking, radiated noise from portable or fixed patient care equipment, equipment service carts, footsteps and dropped objects etc.

   (b) HVAC system sounds transmitted to an enclosed room or open plan space via air supply or return ducts or radiated from unitary equipment such as fan coil units within the space.

   (c) Transmitted sounds from adjacent rooms via common partitions, common suspended ceiling systems or plena, common floor/ceiling systems etc.

   (d) Amplified sounds from paging, and announce systems (wired or wireless).

2.2 Applicable Federal, State and Local Codes and Regulations

Specified acoustical finishes shall conform to all applicable codes and regulations including:

   (1) Standards and details governing other than acoustical properties such as: light reflectance, moisture retention and control, seismic restraint, etc.
   (2) Fire resistance standards
   (3) Special requirements for environmental and bio-hazard safety, etc.

1.6.3.2 Sound & Vibration - Acoustical Finishes and Details

2.3 Design Criteria for Acoustical Finishes

(1) All normally occupied hospital and healthcare facility spaces shall incorporate acoustical finishes to achieve design room average sound absorption coefficient, ($\overline{\alpha}_{design}$), as indicated in Table 2.3-1.[12]

2010 Guidelines Table 1.2-1: Design Room Sound Absorption Coefficients

Table 2.3-1: Recommended design room sound absorption coefficients ($\overline{\alpha}_{design}$)

Space	$\overline{\alpha}$ design	Subjective description
Private patient	0.15	"Average" room
Multi-bed patient	0.15	"Average" room
Corridor	0.15	"Average" room
Waiting area	0.25	"Medium-dry" room
Atrium	0.10	"Medium live" room
Physician's office	0.15	"Average" room
Treatment room	0.15	"Average" room
Additional spaces to be added based on program for particular healthcare facility involved.		

(2) The subjective descriptor corresponding to various values of average room sound absorption coefficients for typical rooms having proportions of about 1 : 1.5 : 2 are indicated in Figure A2.3-1.

Table 2.3-2: Subjective attributes of rooms with various design room sound absorption coefficients, $\overline{\alpha}_{design}$

$\overline{\alpha}$ design	Subjective description
0.40	"dry" room
0.25	"medium-dry" room
0.15	"average" room
0.10	"medium-live" room
0.05	"live" room
0.025 and lower	"very live" room

---

[12] Design room sound absorption coefficients are the sound absorption coefficients for each room surface, averaged over the total room surface area (see Appendix 7.X.A3).

1.6.3.2 Sound & Vibration - Acoustical Finishes and Details

A2.3 Determination of Design Room Average Sound Absorption Coefficient ($\overline{\alpha}_{design}$) and Room Sound Absorption Factors (A_R, sf) [13][14]

By Definition: Design Room Sound Absorption Coefficient ($\overline{\alpha}_{design}$)

$$\overline{\alpha}_{design} = \frac{S_1\alpha_1 + S_2\alpha_2 + ... + S_n\alpha_n}{S_1 + S_2 + ... + S_n}$$

Where: $S_1, S_2, ..., S_n$ = surface areas (sf) of various room finish materials
$\alpha_1, \alpha_2, ..., \alpha_n$ = coefficients of the individual finish materials

Example: A 1000 cu. ft. room (10 ft. x 10 ft. x 10 ft.) finished generally in painted concrete: 6 surfaces at 10 ft. x 10 ft. (4 walls, floor and ceiling). α for painted concrete = 0.01

$$\overline{\alpha}_{design} = \frac{S_1\alpha_1 + S_2\alpha_2 + ... + S_n\alpha_n}{S_1 + S_2 + ... + S_n}$$

$$\overline{\alpha}_{design} = \frac{6(10\,ft. \times 10\,ft.)(0.01)}{600} = .01$$ *Very live room per Table 2.3-2*

Example: Same room as above with ACT ($\alpha = 0.90$) added to ceiling surface:
4 walls and floor, painted concrete, and ceiling, ACT:

$$\overline{\alpha}_{design} = \frac{S_1\alpha_1 + S_2\alpha_2 + ... + S_n\alpha_n}{S_1 + S_2 + ... + S_n}$$

$$\overline{\alpha}_{design} = \frac{5(10\,ft. \times 10\,ft.)(0.01) + (10\,ft. \times 10\,ft.)(0.90)}{600} = 0.16$$ *Average room per Table 2.3-2*

By Definition: Room Sound Absorption Factor (A_R) is given by:

$$A_R = \frac{S\overline{\alpha}}{1-\overline{\alpha}} \cong S_1\alpha_1 + S_2\alpha_2 + ... + S_n\alpha_n \;(sf)$$

Where: S = total room surface area (sf)
$S_1\alpha_1 + S_2\alpha_2 + ... + S_n\alpha_n$ = individual room finish surface areas X their sound absorption coefficients

---

[13] The design room sound absorption coefficient ($\overline{\alpha}_{design}$) is a dimensionless attribute of a room independent of the room's cubic volume and surface area.

[14] The room sound absorption factor (A_R, in sq. ft.) is an attribute of a room of a specific cubic volume and total surface areas made up of various finishes.

1.6.3.2 Sound & Vibration - Acoustical Finishes and Details

Room Sound Absorption Factor (AR) may be determined by executing the above equation for a particular room involved, or may be quickly estimated using Figure A2.3-1:

Figure A2.3-1: Method for determining room sound absorption factor (A_R, sf) from room volume V (cf) and average room sound absorption coefficient ($\overline{\alpha}$) [after Beranek, L.L, *Noise Reduction,* Wiley, 1960]

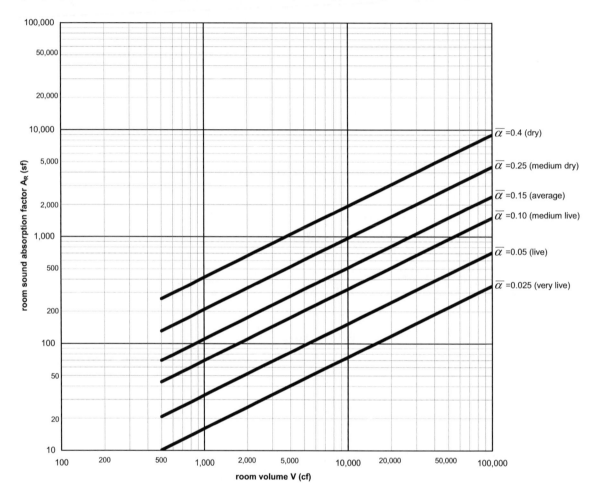

1.6.3.2 Sound & Vibration - Acoustical Finishes and Details

Room sound absorption factor A_R (sf) is an important acoustical parameter in each of the following examples, and may be determined from Figure A2.3-1:

(1) Determining the reduction of "room noisiness" due to sources producing noise within a room to meet predetermined design guidelines.

Example: Determine the reduction in noisiness for a 10,000 cf nurse's station that has been finished with a new acoustical ceiling to change the room from a "live" to an "average" room.

Answer (from Figure A2.3-1):): $A_{R, live}$=160 sf; $A_{R, average}$=500 sf

By definition, noise reduction between two conditions of room absorption:

$$NR = 10\log\left(\frac{A_2}{A_1}\right) = 10\log\left(\frac{500}{160}\right) = 5 dB$$

In other words, the occupants of the acoustically refinished nurse's station will experience a noticeable reduction in room "noisiness" of 5 dB.

(2) Calculating room sound levels due to HVAC noise sources (see Section 3 Room Noise Levels)

Example: Determine room absorption factor A_R for a "medium-live", 2000 cf medical examining room.

Answer (from Figure A2.3-1):): A_R=110 sf

(3) Estimating transmitted sound levels from noise sources in an adjacent room (see Section 4 Sound Isolation Performance of Constructions).

Example: Determine room absorption factor AR for an "average", 1000 cf patient admitting office.

Answer (from Figure A2.3-1): A_R=110 sf

2.4 Considerations for NICUs (from Standard 21: Ceiling Finishes)

Ceilings shall be easily cleanable and constructed in a manner to prohibit the passage of particles from the cavity above the ceiling plane into the clinical environment.

The ceiling construction in infant rooms and adult sleep areas and spaces opening onto them shall not be friable and shall have a noise reduction coefficient (NRC) (22) of at least 0.95 for 80% of the entire surface area or an average of NRC 0.85 for the whole ceiling including solid and acoustically absorptive surfaces. To ensure protection from noise intrusion ceilings in infant rooms and adult sleep areas shall be specified with a ceiling articulation class (CAC) -29. Finishes shall be free of substances known to be teratogenic, mutagenic, carcinogenic, or otherwise harmful to human health.

1.6.3.2 Sound & Vibration - Acoustical Finishes and Details

A2.4.1 Interpretation

Since sound abatement is a high priority in the NICU, acoustical ceiling systems are desirable, but must be selected and designed carefully to meet this standard. In most NICUs the ceiling offers the largest available area for sound absorption. The Standard for ceiling finishes includes areas that communicate with infant rooms and adult sleep areas (e.g., hallways, corridors, storage, and staff work areas) when doors are opened in the course of daily activity.

Ceilings with high acoustical absorption (i.e., high NRC ratings) do not have a significant barrier effect (i.e., offer protection from sounds transmitted between adjacent areas). A CAC-29 provides a moderate barrier effect and allows a broad range of ceiling products. Poor barrier effects can result if room-dividing partitions are discontinued above the ceiling allowing room-to-room cross talk or if there are noise-producing elements in the ceiling plenum. If the ceiling plenum contains noise sources such as fan-powered boxes, in-line exhaust fans, variable air volume devices, etc. then a higher CAC than CAC-29 may be necessary (Philbin & Evans, 2006)

VOCs and PBTs such as cadmium are often found in paints and ceiling tiles and should be avoided. Specify low- or no-VOC paints and coatings.

A2.4.2 Interpretation:

Acoustically absorptive surfaces reduce reverberation and, therefore, sound levels at a distance from the sound source. When possible, two perpendicular walls should be covered with sound absorptive surface materials with an NRC of at least 0.65. Where this is not possible the upper portions of all four walls (above areas likely to be damaged by the movement of equipment) should be covered with such material. Glass should be limited to the area actually required for visualization in order to leave wall surface available for absorptive surface treatment. While a variety of flooring will limit impact noise somewhat, specialized carpeting offers the most protection. Carpeting used in infant areas must have impermeable backing, monolithic or chemically- or heat-welded seams, and be tolerant of heavy cleaning including the use of bleach.

3 ROOM NOISE LEVELS

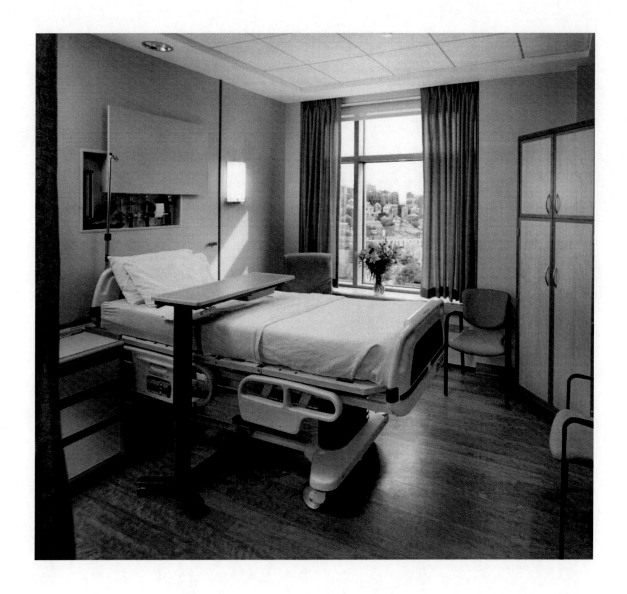

1.6.3.3 Sound & Vibration - Room Noise Levels

3 ROOM NOISE LEVELS

3.1 General

Background sound levels generated by building mechanical systems and other hospital noise sources (MRI, elevators, etc.) shall be considered for all occupied spaces. This section provides design guidelines for achieving appropriate background sound levels for various health care facility spaces.

3.2 Federal, State and Local Codes, Regulations, and Guidelines

(1) Department of Health and Human Services (including HIPAA)
(2) Building code of the local or state jurisdiction
(3) Local planning and zoning limits on building noise.
(4) Occupational Safety and Health Administration (OSHA) – worker noise exposure in areas where sound levels exceed 85 dBA.
(5) AIA Guidelines for noise in NICUs
(6) ASHRAE Applications Handbook

1.6.3.3 Sound & Vibration - Room Noise Levels

3.3 Design Criteria for Room Noise Levels

Noise from building mechanical systems shall be evaluated with one of the following rating systems. One rating system shall be chosen and room noise levels shall fall within the sound level ranges shown for that rating system:[15]

---

[15] Two tables of recommended room noise levels were referenced in the preparation of this guideline.

The first table was taken from the draft standard ANSI S12.2 "Criteria for Evaluating Room Noise".

Room Type	NC or RNC	dBA
Private rooms	25-30	35-39
Wards	30-35	39-44
Operating rooms	25-35	35-44
Laboratories	35-45	44-53
Corridors	35-45	44-53
Public areas	40-45	48-52

The second table was taken from the 2003 ASHRAE Applications Handbook, Chapter 47 "Sound and Vibration Control".

Room Type	RC (N) or NC
Private rooms	25-35
Wards	30-40
Operating rooms	25-35
Corridors and public areas	30-40
Testing/research lab, minimal speech	45-55
Research lab, extensive speech	40-50
Group teaching lab	35-45
Public areas	40-45
Private offices	25-35
Conference rooms	25-35
Teleconference rooms	25 (max)
Large lecture rooms	25-30

RC (N) = Room Criteria with a neutral spectrum

1.6.3.3 Sound & Vibration - Room Noise Levels

2010 Guidelines Table 1.2-2: Minimum-Maximum Design Criteria for Noise in Interior Spaces

Table 3.3-1: Recommended criteria for noise in interior spaces

Room Type	NC / RC(N) /	dBA
Patient rooms	30-40	35-45
Multiple occupant patient care areas	35-45	40-50
NICU[1]	25-35	30-40
Operating rooms[2]	35-45	40-50
Corridors and public spaces	35-45	40-50
Testing/research lab, minimal	45-55	50-60
Research lab, extensive speech[2]	40-50	45-55
Group teaching lab	35-45	40-50
Doctor's offices, exam rooms	30-40	35-45
Conference Rooms	25-35	30-40
Teleconferencing Rooms	25 (max)	30
Auditoria, large lecture rooms	25-30	30-35

Notes:

[1] NICU building mechanical noise levels were set for compliance with AIA requirements when added to NICU activity noise

[2] Noise levels on upper end of range due to practical airflow requirements.

[3] Please see Appendix A3.4.3 for a discussion of different Room Noise Rating Criteria.

NC = Noise Criteria
RC(N) = Room Criteria, Neutral Spectrum
RNC = Room Noise Criterion
dBA = A-weighted Sound Pressure Level

1.6.3.3 Sound & Vibration - Room Noise Levels

3.4 Conformance measurements of room sound level

Room sound levels shall be measured in accordance with applicable ANSI standards in publication.

A3.4.1 Determination of Room Noise Level

The HVAC noise levels in occupied spaces result from a combination of sound transmission paths:

(1) Duct-borne fan noise
(2) Airflow/turbulence noise
(3) Duct breakout noise
(4) Airborne radiated noise through partitions, over demising walls, or through air-transfer ducts
(5) Structure-borne noise from vibration transmission
(6) Duct-borne terminal box noise
(7) Radiated terminal box noise
(8) Acoustical environment (Room Sound Absorption Factor, A_R. See Section 2 Acoustical Finishes and Details)

These paths should all be addressed in order to comply with the design criteria for room sound levels given in Table 3-1. Noise control techniques are discussed in the Sound and Vibration Control chapter of the current ASHRAE Applications Handbook.

A3.4.2 Effect of Background Noise on Clinical Hearing Ability

Groom (American Heart Journal, 1956) studied the effect of background noise on cardiac auscultation. Zun and Downey (Academic Emergency Medicine, 2005) asked clinicians whether they could hear heart and lung sounds from auscultation of a human test subject in the presence of steady 90 dB pink noise. 96% of the clinicians could hear heart sounds and 91% of the clinicians could hear lung sounds. Accuracy of diagnosis was not a part of this study. Studies should be performed on the accuracy of auscultation-based diagnosis in the presence of typical HVAC background sound levels and actual or simulated activity sound levels in modern hospitals.

A3.4.3 Discussion of Background Noise Rating Criteria

The RC Mark II and RNC noise rating criteria methods incorporate sound levels in octave bands below 63 Hz. However, ASHRAE does not provide guidance for prediction of sound levels below 63 Hz. Designers should use industry standard techniques for predicting HVAC noise in the 63 Hz and higher octave bands and should calculate the predicted noise rating criteria based upon the 63 Hz and higher octave bands. Authorities Having Jurisdiction should realize this limitation on the designers' part and should use only the 63 Hz and higher octave bands for evaluation of compliance.

1.6.3.3 Sound & Vibration - Room Noise Levels

ASHRAE prefers the RC Mark II Room Criteria method for the design and evaluation of HVAC noise control. The 2003 ASHRAE Applications Handbook identifies low frequency noise evaluation and sound quality as advantages of RC Mark II over NC. The RC Mark II method uses octave band sound levels from 16 Hz to 4000 Hz provides a means for calculating a sound quality descriptor.

The RC Mark II method is a two-dimensional rating system that has a numerical rating and a sound quality rating. The four sound quality ratings are (N) for neutral, (LF) for low-frequency rumble, (MF) for mid-frequency roar, and (HF) for high-frequency hiss. The neutral (N) sound quality rating is the goal of the RC Mark II method, and it requires that the low-frequency, mid-frequency, and high-frequency sound levels have specific arithmetic relationships to one another to achieve a balanced spectrum. The RC Mark II method does not account for the possibility that low-frequency fan noise levels could change by a different amount than high-frequency turbulence noise during capacity changes in a VAV system, affecting the sound quality rating. The 2007 ASHRAE Fundamentals Handbook states that there is insufficient knowledge to evaluate a RC Mark II rating that has a numerical rating less than the maximum allowable RC Mark II rating but a sound quality rating other than neutral (N). Authorities Having Jurisdiction wishing to use the RC Mark II method for HVAC noise evaluation should require that ASHRAE provide guidance for evaluating RC Mark II ratings with compliant numerical ratings and non-compliant sound quality ratings.

1.6.3.3 Sound & Vibration - Room Noise Levels

1.6.3.3 Sound & Vibration - Room Noise Levels

3.5 Considerations for NICUs (from Standard 23: Acoustic Environment)

Infant rooms (including airborne infection isolation rooms), staff work areas, family areas, and staff lounge and sleeping areas and the spaces opening onto them shall be designed to produce minimal background noise and to contain and absorb much of the transient noise that arises within them.

The combination of continuous background sound and operational sound in infant bed rooms and adult sleep areas shall not exceed an hourly Leq of 45 dB and an hourly L10 of 50 dB, both A-weighted slow response. Transient sounds or Lmax shall not exceed 65 dB, A-weighted, slow response in these rooms/areas. The combination of continuous background sound and operational sound in staff work areas, family areas, and staff lounge areas shall not exceed an hourly Leq of 50 dB and an hourly L10 of 55 dB, both A-weighted slow response. Transient sounds or Lmax shall not exceed 70 dB, A-weighted, slow response in these areas.

To achieve the required noise levels in infant rooms and adult sleep areas building mechanical systems and permanent equipment in those areas shall conform to Noise Criteria (NC) -25 based on manufacturers' noise ratings with allowance for other sound sources and adjustment for room loss if less than 10 dB. Building mechanical systems and permanent equipment in other areas specified in the Standard shall conform to a maximum of NC-30. Building mechanical systems include heating, ventilation, and air conditioning systems (HVAC) and other mechanical systems (e.g., plumbing, electrical, vacuum tube systems, and door mechanisms). Permanent equipment includes refrigerators, freezers, ice machines, storage/supply units, and other large non-medical equipment that is rarely replaced.

A3.5.1 Interpretation:

The permissible noise criteria of an hourly L_{eq} of 45 dB, A-weighted, slow response in infant rooms and adult sleep areas is more likely to be met in the fully operational NICU if building mechanical systems and permanent equipment in those areas and the areas in open communication with them conform to NC-25 or less. NC-25 translates to approximately 35 dBA of facility noise. A realistic addition of 10 dBA of operational noise above this background will result in a L_{eq} of about 45 dBA. Limiting operational noise to only 10 dBA above the background will require conscientious human effort.

A3.5.2 Interpretation:

The type of water supply and faucets in infant areas should be selected so as to minimize noise, and should provide instant warm water in order to minimize time "on".

Noise generating activities (linen and supply carts, conference areas, clerk's areas, multiple-person work stations, and travel paths not essential to infant care),

1.6.3.3 Sound & Vibration - Room Noise Levels

permanent equipment and office equipment should be acoustically isolated from the infant area. Vibration isolation pads are recommended under leveling feet of permanent equipment and appliances in noise-sensitive areas or areas in open or frequent communication with them.

Post-construction validation of specifications for the building mechanical systems and permanent equipment should include noise and vibration measurement, reporting, and remediation. Measurement of NC levels should be made at the location of the infant or adult bed or at the anticipated level of the adult head in other areas. Each bed space must conform to the Standard.

4 SOUND ISOLATION PERFORMANCE OF CONSTRUCTIONS

1.6.3.4 Sound & Vibration - Sound Isolation Performance of Constructions

4 SOUND ISOLATION PERFORMANCE OF CONSTRUCTIONS

4.1 General

(1) Sound isolation shall be considered for all occupied spaces.

(2) Adequate sound isolation will result in speech privacy, acoustic comfort, and a reduction in noise-produced annoyance.

(3) Sound isolation between hospital occupants and noise sources is determined by the sound level difference between source and receiver and by the level of background sound at the receiver's location.

4.2 Applicable Federal, State and Local Codes and Regulation

Specified partitions, floor/ceiling assemblies and related details shall conform to all applicable building codes and regulations, seismic restraint, fire ratings, etc.

4.3 Design guidelines for sound isolation between enclosed rooms

(1) Space layout shall be considered early in the design of health care facility buildings. Non-compatible adjacencies (e.g. noisy equipment spaces or elevator shafts next to patient rooms) shall be avoided to the extent possible to minimize the need for expensive, high-performance demising constructions.

(2) The sound isolation ratings in Table 4-1 are considered the composite sound isolation performance values associated with the demising constructions, whether they are the floor/ceiling or wall partitions.

   (a) Details such as the ceiling plenum conditions, windows, doors, penetrations through the constructions, etc. shall be addressed to provide this composite sound isolation rating.

   (b) Table 4-1 will provide Normal speech privacy (except at corridor walls with doors), assuming a background sound level of at least 30 dBA.

1.6.3.4 Sound & Vibration - Sound Isolation Performance of Constructions

2010 Guidelines Table 1.2-3: Design Criteria For Minimum Sound Isolation Performance Between Enclosed Rooms

Table 4.3-1: Recommended sound isolation performance between enclosed rooms

Adjacency combination		STC_c
Patient Room	Patient Room (horizontal)	45^1
Patient Room	Patient Room (vertical)	50
Patient Room	Corridor (with entrance)	35^2
Patient Room	Public Space	50
Patient Room	Service Area	60^3
Exam Room	Corridor (with entrance)	35^2
Exam Room	Public Space	50
Toilet Room	Public Space	45
Consultation Room	Public Space	50
Consultation Room	Patient Rooms	50
Consultation Room	Corridor (with entrance)	35^2
Patient Room	MRI Room	60^3
Exam Room	MRI Room	60^3
Exam Room	Exam Room (no electronic masking)	50
Exam Room	Exam Room (with electronic masking)	40
Public Space	MRI Room	50

Notes:

[1] In cases where greater speech privacy is required when both patient doors on either side of a patient room wall are closed, the wall performance requirement shall be STC 50.

[2] The performance of this construction assumes a closed door.

[3] STC 60 ratings should be relaxed if in compliance with room noise requirements is achieved with lower performance constructions. See Table 3.3-1.

(3) Sound transmission class (STC) performance shall be determined by tests in accordance with methods set forth in ASTM E90 and ASTM E413 (current description from AIA HCFGL).

(4) Where partitions do not extend to the structure above, sound transmission through ceilings and composite STC performance must be considered. Demountable wall STC performance shall be balanced with the Ceiling Attenuation Class (CAC) performance of the ceiling.

(5) Public Space includes corridors (except patient room access corridors), lobbies, dining rooms, recreation rooms, and similar space that is not directly associated with the healthcare mission.

1.6.3.4 Sound & Vibration - Sound Isolation Performance of Constructions

(6) Service areas include kitchens, elevators, elevator machine rooms, laundries, garages, maintenance rooms, boiler and mechanical equipment rooms, and similar non-public spaces that serve the operation of the facility.

(7) Structure-borne sound transmission shall be addressed in accordance with Section 6, Building Vibration.

(8) If the ratio of the partition surface area to the room sound absorption factor is greater than 1, the quantity 10*log (partition surface area/room sound absorption factor) should be added to the STC requirement. Room sound absorption factor shall be calculated according to Section A2.3.

(9) Table A4.3.3-1 shows that open doors severely degrade the performance of walls.

(10) STC is a laboratory rating and cannot be measured in the field. Apparent STC (ASTC) is a field measurement described in ASTM E336-05. This standard states "the actual transmission loss of the partition will usually be higher than the apparent transmission loss".

(11) MRI noise and vibration transmission to nearby sensitive spaces requires special consideration.

A4.3.1 Typical partitions

(1) Floor-ceiling partition: STC50
- *5-inch average thickness normal weight concrete slab*
- *suspended acoustical tile ceiling below*

(2) Wall partition: STC40
- *(1) layer 5/8" GWB each side*
- *metal stud*
- *no insulation*

(3) Wall partition: STC45
- *(1) layer 5/8" GWB each side*
- *metal stud*
- *insulation in cavity*

(4) Wall partition: STC50
- *(2) layers 5/8" GWB each side*
- *metal stud*
- *insulation in cavity*

A4.3.2 Extraordinary partitions

(1) Floor-ceiling partition: STC60
- *5-inch average thickness normal weight concrete slab*
- *floated 4-inch concrete floor slab on glass fiber or neoprene mounts above*

(2) Wall partition: STC60

1.6.3.4 Sound & Vibration - Sound Isolation Performance of Constructions

- *Six-inch painted CMU block*
- *1-inch of space between the block and stud framing*
- *metal studs with batt insulation in the cavity*
- *two layers of gypsum board on the occupied side of the metal studs*

1.6.3.4 Sound & Vibration - Sound Isolation Performance of Constructions

A4.3.3 Composite Sound Transmission Class (STCc)

The following table can be used to estimate the composite STC rating (STCc) of interior partitions. First, determine which typical wall type will be used. Calculate the percentage of the wall occupied by a window or door. Read the STCc from the intersection of the door/window row and typical wall column. A generalized graph illustrating this calculation can be found on page 191 of Egan, Architectural Acoustics, 1988.

Table A4.3.3-1: Composite Sound Transmission Class ratings for various partition systems

Weaker element (% wall area)	STC of door or window	STC 40 (1) layer 5/8" GWB each side, metal stud	STC 45 (1) layer 5/8" GWB each side, metal stud, with insulation	STC 50 (2) layers 5/8" GWB each side, metal stud, with insulation
Open door (10%)	0	10	10	10
Open door (20%)	0	7	7	7
Un-gasketed door (10%)	20	30	30	30
Un-gasketed door (20%)	20	27	27	27
1/8-inch glazing (10%)	26	35	36	36
1/8-inch glazing (20%)	26	33	33	33
Gasketed door or 1/4-inch glazing (10%)	30	37	39	40
Gasketed door or 1/4-inch glazing (20%)	30	35	36	37
1/4-inch laminated glazing (10%)	35	39	42	44
1/4-inch laminated glazing (20%)	35	39	40	41
Acoustically rated STC-45 door or insulating glazing with 4-inch airspace (10%)	45	40	45	50
Acoustically rated STC-45 door or insulating glazing with 4-inch airspace (20%)	45	40	45	49

1.6.3.4 Sound & Vibration - Sound Isolation Performance of Constructions

4.4 Design guidelines for speech privacy between enclosed rooms

Protecting the privacy of speech in healthcare facilities reduces medical errors by assuring open dialogue between patients, families, practitioners and administrative staff. Speech privacy is also required under HIPAA and other federal, state and local privacy protection statutes, and is included as a performance parameter in inspections by JCAHO, FDA and others and is included in the satisfaction indices of organizations such as Press Gainey.

Speech privacy is defined by ANSI T1.523-2001/Glossary as "Techniques to render speech unintelligible to casual listeners" and is similarly defined by other standards (i.e. ANSI S3.5-1969 & 1997). Further, ASTM E1130 (R1997 & R2006) and other standards specify consistent, measurable, numeric levels of speech privacy, i.e., "Normal Privacy" versus "Confidential Privacy", as well as best practices for achieving them. The standards further describe objective and quantitative methods and procedures, available equipment and best practices that are to be used for measuring, monitoring, mitigating and certifying speech privacy conditions. Based on six decades of research and practice, these speech privacy standards are long-standing, consistent with each other, and widely used by engineering professionals.

(1) Speech privacy is a critical element to achieve for the patients' rights to privacy of oral communication per HIPAA and the functional program.

(2) Speech privacy shall be measured in terms of the Articulation Index (AI), Privacy Index (PI), Speech Transmission Index (STI) or Speech Intelligibility Index (SII).

2010 Guidelines Table 1.2-4: Design Criteria for Speech Privacy for Enclosed Room and Open-Plan Spaces

Table 4.4-1: Speech privacy goals for enclosed rooms

Enclosed Rooms Goal	AI	PI	STI	SII
Normal	≤0.15	≥85%	≤0.19	≤0.20
Confidential	≤0.05	≥95%	≤0.12	≤0.10
Secure	Special consideration required.			

Table 4.4-2: Speech privacy goals for open-plan spaces

Open Plan Goal	AI	PI	STI	SII
Normal (non-intrusive)	≤0.20	≥80%	≤0.23	≤0.25
Confidential	Special consideration required.			

1.6.3.4 Sound & Vibration - Sound Isolation Performance of Constructions

(3) Normal speech privacy shall generally be provided between enclosed rooms.

(4) Confidential speech privacy shall be achieved in admitting areas, areas where patients discuss their personal health, psychiatric and psychological testing rooms, hematology labs, exam rooms, etc.

(5) Secure speech privacy is the highest order and generally requires total inaudibility of speech in adjacent spaces. This necessitates extensive analysis of the uses of the space, including anticipated speaker effort, constructions separating the spaces, and background sound levels.

A4.4 Enclosed Room Speech Privacy Design Guidance

(1) To achieve confidential speech privacy the sum of the composite STC and the A-weighted background noise level shall be at least 75.

(2) Speech privacy can be achieved with proper space planning, partitions, room finishes and effective use of sound masking systems.

4.5 Design guidelines for speech privacy in open-plan spaces

(1) People working in open plan spaces are most productive when there is a minimum of distraction from voices, equipment, etc. This involves designing the acoustical environment to minimize such distractions and improve a person's ability to concentrate (See Section 4.4 for design criteria for areas that require increased privacy to comply with HIPAA guidelines for speech privacy).

(2) Confidential speech privacy is not readily achievable in open plan spaces due to the lack of barriers, low ambient sound levels and typical voice effort.

(3) Options for achieving confidential speech privacy in open plan spaces could include having a separate room where these conversations can occur in private.

(4) PI (Privacy Index) criteria are defined in ASTM standard E1130.

(5) SII (Speech Intelligibility Index), defined in ANSI standard S3.5-1997, replaces the AI (Articulation Index) of ANSI standard S3.5-1969.

(6) The STI (Speech Transmission Index) is defined in IEC 60268-16. The STI metric is provided in these criteria since it is a relatively easy measurable criterion, which can be determined by several instruments that are commercially available.

(7) The SII criterion can be used in the design of the open office space; it can also be used to quantitatively assess field conditions, though it requires more extensive calculations to determine the results.

4.6 Open Plan Speech Privacy Design Guidance

(1) To achieve confidential speech privacy the sum of the composite STC and the A-weighted background noise level shall be at least 75.

(2) Speech privacy can be achieved with proper space planning, barriers, room finishes and effective use of sound masking systems.

1.6.3.4 Sound & Vibration - Sound Isolation Performance of Constructions

4.6.1 Considerations for NICUs (from Standard 23: Acoustic Environment)

Speech privacy and freedom from intrusive sounds shall be provided by acoustic seals for doors and building to meet STC criteria (below) for demising partitions in infant rooms, on-call and sleep rooms, family transition rooms, and conference rooms or offices in which sensitive staff and family information is discussed. All other penetrations for conduits, inset boxes, pipes, ducts, and other elements in sound demising partitions shall be sealed airtight to prevent noise flanking (leakage) through gaps and openings.

4.6.2 Considerations for NICUs (from Standard 23: Acoustic Environment)

With space at a premium, many incompatible adjacencies are possible in NICU designs (e.g., break area, meeting room, or mechanical room sharing a wall with an infant or adult sleep room). Specialized wall and floor/ceiling treatments will help to meet criteria in these non-optimal conditions.

The criteria below are for sound transmission loss (TL) or attenuation through horizontal barriers (e.g., walls, doors, windows) and vertical barriers (e.g., between floors). The Sound Transmission Class (STC) rating spans speech frequencies and is relevant for separation of spaces with conversational and other occupant-generated noise. The Noise Reduction (NR) rating, which covers a wider frequency span, is more relevant for mechanical noise dominated by low frequencies. The recommended criteria for TL below apply to barriers between adjacent spaces and infant areas or adult rest or sleep rooms.

2010 Guidelines Table A2.1-a: Sound Transmission Loss or Attenuation Through Horizontal and Vertical Areas in NICUs

Adjacency combination		STC_c
NICU	Pedestrian-only corridor	45
NICU	Equipment corridor	55
NICU	Infant area	40
NICU	Reception	55
NICU	Meeting room with amplified	55
NICU	Staff work area	55
NICU	Administrative office,	45
NICU	Non-related area	50
NICU	Mechanical area	NR 60-65
NICU	Electrical area	NR 50-55

Adapted from Evans JB, Philbin MK. Facility and operations planning for quiet hospital nurseries. J Perinatol 2000; 20(8):S105-12. Revised and reprinted with permission of Jack B. Evans, PE, M. Kathleen Philbin, RN, PhD, The Journal of Perinatology, and Nature Publishing Company).

1.6.3.4 Sound & Vibration - Sound Isolation Performance of Constructions

Sound transmission from the exterior of the building should meet the NC criteria inside all spaces identified in the Standard.

It is advisable to enlist the services of an acoustical engineer from the onset of the project through post-construction validation. This specialty service is usually not covered by architectural fees and can assist in program and design development, design of mechanical systems, specification of equipment and building construction, and test and balance validation. Enlistment of acoustical services late in the design process often results in fewer and more costly options for meeting performance standards.

Speech Privacy References:

(1) ANSI S3.5-1997, "Methods for Calculation of the Speech Intelligibility Index", American National Standard, Standards Secretariat, Acoustical Society of America, New York, USA.

(2) IEC 60268-16:2003(E), "Objective rating of speech intelligibility by speech transmission index", International Electrotechnical Commission.

(3) ASTM E1130 "Standard Test Method for Objective Measurement of Speech Privacy in Open Offices Using Articulation Index".

(4) "Validation of Architectural Speech Security Results", Bradley, J.S.; Gover, B.N., IRC-RR-221, National Research Council Canada, March 2006.

(5) "Masking Speech in Open-plan Offices with Simulated Ventilation Noise: Noise Level and Spectral Composition Effects on Acoustic Satisfaction", Veitch, J.A., Bradley, J.S.; Legault, L.M., Norcross, S., and Svec, J.M., IRC-IR-846, National Research Council Canada, April 2002.

(6) "Describing Levels of Speech Privacy in Open-plan Offices", Bradley, J.S.; Gover, B.N., IRC-RR-138, National Research Council Canada, 12 September 2003.

(7) "Assessment of Architectural speech security of closed offices and meeting rooms". Bradley, J.S.; Gover, B.N. The Journal of the Acoustical Society of America -- October 2004 -- Volume 116, Issue 4, p. 2612.

(8) "Masking Speech in Open-plan Offices with Simulated Ventilation Noise: Noise Level and Spectral Composition Effects on Acoustic Satisfaction", Veitch, J.A., Bradley, J.S.; Legault, L.M., Norcross, S., and Svec, J.M., IRC-IR-846, National Research Council Canada, April 2002.

A4.6 Interpretation:

The acoustic environment is a function of both the facility (e.g., building mechanical systems and permanent equipment, the intrusion of exterior sounds, the sound containment afforded by doors and walls, and the sound absorption afforded by surface finishes) and operations (e.g., the activities of people and function of medical equipment and furnishings).

1.6.3.4 Sound & Vibration - Sound Isolation Performance of Constructions

The acoustic conditions of the NICU should favor speech intelligibility, normal or relaxed vocal effort, speech privacy for staff and parents, and physiologic stability, uninterrupted sleep, and freedom from acoustic distraction for infants and adults [24]. Such favorable conditions encompass more than the absence of noise and require specific planning for their achievement. Speech Intelligibility ratings in infant areas, parent areas, and staff work areas should be "good" to "excellent" as defined by the International Organization for Standardization ISO 9921:2003. Speech intelligibility for non-native but fluent speakers and listeners of a second language requires a 4 to 5 dBA improvement in signal-to-noise ratio for similar intelligibility with native speakers. The L_{eq}, L_{10} and L_{max} limits will safeguard this intelligibility and also protect infant sleep [25].

5 PAGING & CALL SYSTEMS, CLINICAL ALARMS, MASKING SYSTEMS & SOUND REINFORCEMENT

1.6.3.5 Sound & Vibration - Paging, Alarm, Masking & Sound Reinforcement Systems

5 PAGING & CALL SYSTEMS, CLINICAL ALARMS, MASKING SYSTEMS & SOUND REINFORCEMENT

5.1 General

Electro-acoustic systems can affect the acoustical environment of healthcare facilities, and the acoustical environment can affect the perception of these systems. Patient safety and comfort, as well as staff comfort and productivity, are considerations in the configuration of these systems.

5.2 Applicable Federal, State and Local Codes and Regulation

(1) Department of Health and Human Services (including HIPAA)
(2) Building code of the local or state jurisdiction
(3) National Fire Protection Association (NFPA)
(4) Joint Commission on Accreditation of Healthcare Organizations (JCAHO)

5.3 Paging and Call Systems

(1) Voice paging and call systems shall be designed to achieve a minimum Speech Transmission Index (STI) of 0.50 or a Common Intelligibility Scale (CIS) rating of 0.70 at representative points within the area of coverage to provide acceptable intelligibility from the system. The conversion between CIS and other scales of intelligibility is available from Annexes A and B of IEC 60489-Sound Systems for Emergency Purposes (NFPA 72-2002).

1.6.3.5 Sound & Vibration - Paging, Alarm, Masking & Sound Reinforcement Systems

(2) Performance of the system shall achieve:

  (a) 70 dBA minimum sound level; or

  (b) 10 dBA above background noise levels (whichever is higher); and

  (c) coverage within +/- 4 dB at the 2000 Hz octave band throughout corridors, open treatment areas and public spaces .

(3) Wireless communication devices such as IP phones, wearable communication badges, and vibrating beepers shall be considered as options to notify clinical staff and reduce the use of overhead paging systems.

(4) Wireless asset tracking technologies such as RFID and infrared shall be considered as options for staff, patient, and equipment location to reduce the use of overhead paging systems.

(5) Integration of call systems with these wireless communication and location devices shall also be considered.

5.4 Clinical alarms

Clinical alarms shall be audible according to ISO 7731 "Danger signals for work places – Auditory danger signals".[16][17]

---

[16] Use of the effective masked threshold technique, while more complicated than an overall A-weighted signal-to-noise ratio, allows for lower alarm sound levels while maintaining audibility.

[17] ISO 7731 states: "The sound pressure level of the alarm signal in one or more 1/3 octave bands should exceed the effective masked threshold by 13 dB in the 1/3 octave band under consideration." (There is also a criterion based on octave band measurements.) Section 5.1 of ISO 7731 also calls for the measurements of the background sound and alarm sound levels to be the maximum readings with time weighting "Slow".

1.6.3.5 Sound & Vibration - Paging, Alarm, Masking & Sound Reinforcement Systems

A5.4 Audibility of tonal alarms

(1) The 2002 edition of NFPA 72, the National Fire Alarm Code, provides a method for measuring the audibility of narrow band tonal alarms using the techniques in ISO 7731. These techniques use the favorable audibility of tonal sounds versus broadband sounds in the midst of competing noise, based on staff training.[18]

(2) Where possible, clinical alarms should be assessed to confirm whether sound levels can be reduced for patient comfort.[19]

5.5 Masking Systems

(1) Sound masking systems are useful tools for reducing patient distractions and enhancing speech privacy in all types of medical facilities.

(2) Systems shall be designed for levels that do not exceed 48 dBA.

(3) Loudspeaker coverage shall provide for uniformity of +/- 2 dBA.

(4) Suitable spectra shall be designed to effectively mask speech spectra, as described in Reference 1 below.

5.6 Sound Reinforcement

(1) All large conference rooms and auditoria seating more than 25 persons shall consider sound reinforcement and AV playback capabilities.

---

[18] The proper signal-to-noise ratio of alarm sound to effective masked threshold for clinical alarms is not clear at the time of this writing. Neither is the type of background sound measurement necessary to calculate the effective masked thresholds. Ideally, the alarm level would change to track changing background sound levels. In the absence of this technology, one background measurement will be required to determine audibility. Perhaps the reading should be a 60-second Leq taken during a shift change. The effective masked threshold for each 1/3 octave frequency band should be calculated from this Leq measurement in accordance with Annex B of ISO 7731.

[19] Regarding hospital staff and fire alarm audibility thresholds, the NFPA 72 Annex states "lower audible levels are permitted because part of the staff's job is to listen for and respond appropriately to the fire alarm signals." This statement could inform the setting of clinical alarm thresholds.

1.6.3.5 Sound & Vibration - Paging, Alarm, Masking & Sound Reinforcement Systems

(2) Sound reinforcement system shall achieve a minimum Speech Transmission Index (STI) of 0.60 or a Common Intelligibility Scale (CIS) rating of 0.77 at representative points within the area of coverage to provide acceptable intelligibility from the system.

(3) Performance of the system shall achieve:

  (a) 70 dBA minimum sound level; or

  (b) 10 dBA above background noise levels (whichever is higher); and

  (c) coverage within +/- 3 dB at the 2000 Hz octave band throughout the space.

(4) Upgraded sound isolation shall be considered for acoustically sensitive spaces that are adjacent to spaces with sound reinforcement systems.

5.7 Considerations for NICUs (from Standard 23: Acoustic Environment)

Where personal address speakers are located in sensitive areas, announcing systems shall have adjustable volume controls for the speakers in each room and for each microphone that sends signal through the system.

A5.7 Interpretation:

Fire alarms in the infant area should be restricted to flashing lights without an audible signal. The audible alarm level in other occupied areas must be adjustable. Telephones audible from the infant area should have adjustable announcing signals.

References

(1) "Masking Speech in Open-plan Offices with Simulated Ventilation Noise: Noise Level and Spectral Composition Effects on Acoustic Satisfaction", Veitch, J.A., Bradley, J.S.; Legault, L.M., Norcross, S., and Svec, J.M., IRC-IR-846, National Research Council Canada, April 2002.

(2) NFPA 72-2002

6 BUILDING VIBRATION

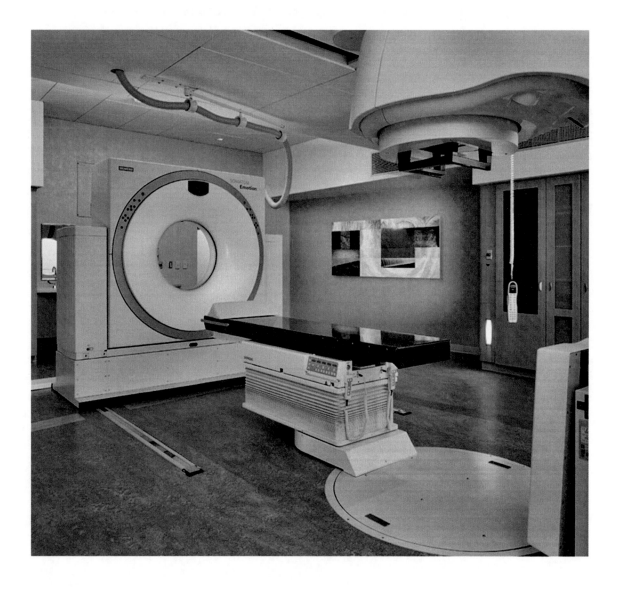

1.6.3.6 Sound & Vibration - Building Vibration

6 BUILDING VIBRATION

6.1 General

Building vibration as it is used here is usually vertical vibration produced by building equipment and activities. It is not vibration produced by earthquakes which is addressed under structural sections of this guide.

Vibration produced by building mechanical equipment, footfall, road and or rail traffic, and medical equipment shall be considered in the design of a hospital building. Seismic restraint is covered elsewhere in the AIA Design Guidelines. Seismic restraint shall be compatible with vibration isolation methods covered in this section.

(1) Mechanical equipment includes all equipment in the construction contract and equipment for which provisions are being made, even if it is not included in the contract.

(2) Often the most significant source of vibration at above grade building locations is footfall.

(3) Road and rail traffic are normally located outside the building footprint. However, parking garages and roadways can be part of the building structure, or pass through them. The vibration that they may produce can affect occupied areas.

(4) Certain types of medical equipment are sensitive to vibration. These include surgical microscopes, electron microscopes, bench microscopes, MRI systems, and various scanning systems.

(5) Certain medical equipment can also produce vibration that can interfere with nearby occupancies in a building, for example MRIs and compressors that are part of certain instrumentation. Gurneys with out-of-round wheels or wheels fouled with embedded scrap can also cause excessive floor vibration.

(6) Building floor discontinuities such as expansion breaks and transitions between floor finish types should be minimized to avoid impacts when engaged by rolling equipment.

1.6.3.6 Sound & Vibration - Building Vibration

6.2 Applicable Federal, State and Local Codes and Design Guides

(1) Federal state and local codes and guidelines vary between jurisdictions. Some codes set limits on vibration, such as that produced by blasting construction and industrial activity.

(2) ANSI S2.71-1983 (R2006) Guide to the Evaluation of Human Exposure to Vibration in Buildings

(3) Current ASHRAE Applications Handbook

(4) AISC Design Guide 11 – Floor Vibration due to Human Activity [Note: this guide contains some errors, including in the criteria table, and needs revision. An ASCE paper updating it is in preparation..]

(5) Medical and laboratory instrumentation supplier criteria

6.3 Vibration Control and Isolation

Vibration levels in the building shall not exceed applicable guidelines and limits.

6.3.1 Mechanical, Electrical and Plumbing Equipment (MEP)

(1) All rotating or vibrating building mechanical, electrical, and plumbing (MEP) equipment shall be mounted on vibration isolators.[20]

(2) Bases and supports shall be provided as needed to facilitate attachment of vibration isolators to equipment items.[21] The types of isolators and isolator static deflections shall be as recommended in the most current ASHRAE Applications Handbook.

(3) Elevator type, location, and surrounding structure must be investigated for transmission of vibration and structureborne sound.

6.3.2 Structural

(1) Footfall vibration in the building structure shall be evaluated using American Institute of Steel Construction (AISC) Design Guide 11 – Floor Vibration due to Human Activity.

(2) The structural floor shall be designed to avoid footfall vibration levels not to exceed the peak vibration velocities in Table 6.3.2-1.

(3) Lower levels may be required in order to avoid interference with medical and laboratory instrumentation. See 6.3.4 Medical and laboratory instrumentation.

---

[20] See ASHRAE Chapter 47 for descriptions of vibration isolators and the use of various types.

[21] Consult equipment manufacturer on proper base type.

1.6.3.6 Sound & Vibration - Building Vibration

2010 Guidelines Table 1.2-5: Maximum Limits on Footfall Vibration in Health Care Facilities

Table 6.3.2-1: Recommended limits on footfall vibration in hospitals.

Space Type	Footfall Vibration Peak Velocity (μin/s)
Patient Rooms and other Patient Areas	4000[1]
Operating and other Treatment Rooms	4000[2]
Administrative Areas	8000
Public Circulation	8000

Notes:

1 ANSI S3.29 combined threshold of perception rms velocity times 1.4

2 ANSI S3.29 combined threshold of perception rms velocity.

6.3.3 Structureborne sound

Structureborne sound is vibration transmitted from a source through the building structure and re-radiated as sound at some other and sometimes remote location. Structureborne sound is usually controlled by vibration isolating the source from the building.

(1) Structure-borne transmitted sound shall not exceed the limits for airborne sound presented in Section 3 - Room Noise Levels.

(2) Where necessary, vibration isolate potential sources of structureborne sound.

6.3.4 Medical and laboratory instrumentation

(1) Footfall and equipment vibration velocities on floors shall not exceed the limits recommended by suppliers of the equipment to be housed on these floors.

(2)

(3) In absence of criteria provided by the equipment suppliers or if the equipment to be installed is not known, the velocities on the floors should not exceed the values given in AISC Design Guide 11.

GLOSSARY

1.6.3 Sound & Vibration - Glossary, Abbreviations & References

GLOSSARY

The definitions of acoustical terms used in this publication are most often based on ANSI S1.1-1994 American National Standards Acoustical Terminology. Some of the acoustical terms briefly defined below are explained in greater detail elsewhere in acoustical texts and design manuals.

A-Weighting (dBA)
The filtering of sound that replicates the human hearing frequency response. The human ear is most sensitive to sound at mid frequencies (500 to 4,000 Hz) and is progressively less sensitive to sound at frequencies above and below this range. A-weighted sound level is the most commonly used descriptor to quantify the relative loudness of various types of sounds with similar or differing frequency characteristics.

Absorption
The attenuation (or reduction) of sound level that results when sound propagates through a medium (usually air) or through a dissipative material (sound absorptive material) such as glass fiber or open-cell foam. In the case of sound absorptive materials used in the building industry, attenuation of sound is produced by the conversion of molecular motion, which is sound, into thermal energy due to friction of air molecules with fibrous or cellular materials.

Acoustics
(1) Acoustics is the science of sound, including its production, transmission and effects.

(2) The acoustics of a room are those qualities that together determine its character with respect to the perception of sound.

Acoustical Double Glazing
Two monolithic glass panels, set into a frame, with an air space between the two panels that is usually larger than 1" and most often not hermetically sealed.

Sound and Vibration Design Guidelines for Hospital and Healthcare Facilities

1.6.3 Sound & Vibration - Glossary, Abbreviations & References

Adult sleep areas
Rooms designated for patent or staff sleep or rest.

Ambient Noise
Ambient noise encompasses all sound present in a given environment, being usually a composite of sounds from many sources near and far.

Areas in open acoustic communication
Areas without a barrier wall or an operable door between them or areas separated by a door that is intended to remain open most of the time.

Background or facility noise
Background noise refers to the continuous ambient sound in a space owing to the mechanical and electrical systems of the facility or building itself and to permanent equipment.

Band Pass Filter
The filtering of sound within specified frequency limits or frequency bands. The audible frequency range is often sub-divided into octave, one-third octave, or other fractions of octave bands.

Barriers
A solid obstacle that blocks the line-of-sight between a sound source and a receiver, thereby providing barrier attenuation, i.e., reducing sound level at the receptor. Sound attenuation provided by barriers is related to the transmission loss through the barrier material and diffraction of sound over and around the barrier.

CAC (Ceiling Attenuation Class)
A laboratory-derived rating for the sound attenuation of a suspended ceiling system over two adjacent rooms sharing a common plenum.

Ceiling plenum
The area between the finished ceiling and the underside of the structure above, often used for ductwork, electrical wiring, plumbing pipes, etc. as well as for recessed ceiling lights.

Coincidence Effect/Critical Frequency
A significant reduction in sound transmission loss (i.e., a significant increase in the transmission of sound) through a partition that occurs at critical frequency. The critical frequency is the frequency at which the wavelength of sound in air equals the flexural bending wavelength in the partition or material.

Damping
Damping is the dissipation of vibratory energy in solid media and structures with time or distance. It is analogous to the absorption of sound in air.

1.6.3 Sound & Vibration – Glossary, Abbreviations & References

Day Night Sound Level (DNL, L_{dn})
The 24-hour energy average sound level where a 10 dB "penalty" is applied to sound occurring at night between 10:00 PM and 7:00 AM. The 10 dB penalty is intended to account for the increased sensitivity of a community to sound occurring at night.

Decibel (dB)
A dimensionless unit which denotes the ratio between two quantities that are proportional to power, energy, or intensity. One of these quantities is a designated reference by which all other quantities of identical units are divided. The sound pressure level in decibels is equal to 10 times the logarithm (to the base 10) of the ratio between the pressure squared divided by the reference pressure squared. The reference pressure used in acoustics is 20 microPascals.

Demising partitions
A "demising" assembly, partition, floor, ceiling, etc. is one that separates two different occupants, departments, or functional spaces.

Double Laminated Insulating Glazing
Two laminated glass panels set into a frame that provides an air space between the two glass panels. Such units may or may not be hermetically sealed and can have varying air space thicknesses, depending on acoustical/thermal requirements.

Energy Average Sound Level
Typically, in real world circumstances, sound levels vary considerably over time. The Leq is energy average sound level over a monitoring time interval. It is a hypothetical continuous sound level that contains the same sound energy as the actual sound level occurring during the time interval. The letter symbol L_{eq} typically implies A-weighting, i.e., the energy average sound level in dBA. Also the duration of measurement is typically stated, e. g., L_{eq}(1 hour).

Field Sound Transmission Class (FSTC)
A rating of the field-derived airborne sound transmission loss data for a structure determined in accordance with the procedure of ASTM e336 and E413. The same as STC rating (see below) except as measured in the field in accordance with standard methods. The FSTC is used to quantify actual as-built partition transmission loss across the frequency range of speech sounds. The FSTC incorporates corrections for receiving room sound absorption, and requirements to assess and eliminate sound flanking paths. Refer to ASTM E336, "Standard Test Method for Measurement of Airborne Sound Insulation in Buildings," and ASTM E413, "Classification for Rating Sound Insulation."

Flanking
The transmission of sound around the perimeter or through holes within partitions (or barriers) that reduces the otherwise obtainable sound transmission loss of a partition. Examples of flanking paths within buildings are ceiling plena above partitions;

1.6.3 Sound & Vibration - Glossary, Abbreviations & References

ductwork, piping, and electrical conduit penetrations through partitions; back to back electrical boxes within partitions, window mullions, etc.

Frequency
Frequency is the number of oscillations or cycles per unit time. In acoustics, frequency usually is expressed in units of Hertz (Hz), where one Hertz is equal to one cycle per second.

Fundamental Frequency
The fundamental frequency of an oscillating system is the lowest natural frequency of that system.

Glazing
Glass and glazing components such as interlayer, desiccant, frames, air space, etc., that are installed into a window frame.

Interlayer
The transparent damping material used in laminated glass.

Laminated Glass
A glazing panel composed of two or more panels of monolithic glass panel separated by a transparent damping material.

Laminated Insulating Glazing
A laminated glass panel and a monolithic glass panel set into a frame that provides an air space between the two glass panels.

Mass
Mass is the fundamental property of a material relevant to sound transmission loss through that material. Generally, the more massive the material, the greater the sound transmission loss.

Mass Law Sound Transmission Loss
Below about half the critical frequency, sound transmission loss is generally only related to the mass of a material or partition. Mass Law helps quantify the sound transmission loss at these frequencies. At these frequencies, doubling the mass per unit area of a partition panel, or doubling the frequency for a given mass per unit area, increases the sound transmission loss by 6 decibels in the frequencies controlled by Mass Law.

Monolithic Glass
Glass having a single uniform thickness.

1.6.3 Sound & Vibration – Glossary, Abbreviations & References

Noise

(1) Noise is any undesired sound. By extension, noise is any unwanted disturbance within a useful frequency band, such as excessive traffic sound transmission into a sensitive building space.

(2) Noise is an erratic, intermittent or statistically random oscillation.

Noise Criterion Curves (NC)

A set of approximate equal loudness curves used to assess the acceptability of background sound in buildings. The curve shapes are set to be monotonic in shape and to have loudness levels in phons that are 22 units above the corresponding SIL values. The curves are used along with a measured or estimated octave band spectrum to determine the NC rating of the spectrum. Criteria for acceptable sound in buildings are often expressed as ranges of acceptable NC ratings. NC criteria are generally applied to background sound in buildings produced by building heating, ventilating, and air conditioning (HVAC) equipment. The NC rating of sound produced by an HVAC noise source involves overlaying the measured or estimated octave band spectrum on a series of Noise Criteria curves. The highest NC curve tangent to the measured spectrum in any octave band is the NC rating of that spectrum. For a further explanation, please see American National Standard ANSI S12.2 Criteria for Evaluating Room Noise and ASHRAE 2003 HVAC Applications Handbook, Chapter 47.

Noise Reduction between Rooms

The arithmetic difference between the sound level in a source room and the sound level produced by that source in an adjacent receiving room. The Noise Reduction (NR) is expressed in decibels.

Octave

The ratio of a higher and lower frequencies that equals two.

Octave Band

Groups of frequencies defined by standards where the upper frequency of each band is equal to twice the lower frequency of each band. Octave bands are usually named by their geometric center frequency. For example, the octave band extending between 44.7 Hz and 89.1 Hz is called the 63 Hz octave band. The octave band extending between 89.1 Hz and 178 Hz is called the 125 Hz octave band. The full complement of octave bands in the audible frequency range is as follows: 31, 63, 125, 250, 500, 1000, 2000, 4000, 8000, and 16,000 Hz.

Octave Band Sound Pressure Level

Sound pressure level for all sound contained within a specified octave band.

Oscillation

In acoustics, pressure oscillation is the variation of sound pressure with time alternately above and below the ambient static pressure.

1.6.3 Sound & Vibration - Glossary, Abbreviations & References

Permanent Equipment
Large equipment that is necessary for essential functions of the facility and that is rarely replaced. Such equipment includes refrigerators, freezers, ice machines, mechanical/electrical storage systems for supplies and medication. Permanent equipment is distinct from medical equipment used for direct patient care.

Pitch
That attribute of auditory sensation expressed in terms of sounds being ordered on a scale extending from low to high. Pitch depends primarily upon the frequency of the sound stimulus.

Residual Noise
The lowest levels of sound reached during a monitoring interval, usually produced by distant traffic or industrial activity.

Reverberation
(1) Reverberation is the persistence of sound in an enclosed space resulting from multiple reflections after a sound source has stopped.

(2) Reverberation is the sound that persists in an enclosed space, as a result of repeated reflection or scattering, after the source of the sound has stopped.

Reverberation Time
The reverberation time of a room is the time it takes for sound to decay by 60 dB once the source of sound has stopped.

Room Criteria Curves (RC)
A set of contours that serve as optimum spectrum shapes for background sound in buildings. Octave band spectra that align with a single RC curve are considered neutrally balanced. Neutrally balanced spectra have the proper amounts of low, mid, and high frequency sound energy to cause them to be perceived as innocuous, even though audible. RC curves are straight lines set at -5 dB/octave slopes. The RC rating of a spectrum is determined through methods defined in standards in the ASHRAE Guide. The rating method arrives at a single-number rating with other designators identifying neutral, rumbly, or hissy characteristics. For a further explanation, please see ASHRAE 2003 HVAC Applications Handbook, Chapter 47.

Room Noise Criteria (RNC)
A set of approximate equal loudness curves, similar to the NC curves, used to assess the acceptability of background sound in buildings. Like the NC rating, the RNC rating is the highest curve tangent to a measured or calculated sound level spectrum overlain on a RNC curve set. However, the RNC method also incorporates an adjustment to its rating method to account for the time varying characteristics of sound sometimes experienced in large ventilation systems. For a further explanation, please see American National Standard ANSI S12.2 Criteria for Evaluating Room Noise.

1.6.3 Sound & Vibration - Glossary, Abbreviations & References

Sabin
A unit of absorption expressed in square feet (English units) or square meters (metric units).

Sound
(1) Sound is an oscillation in pressure, stress, particle displacement, particle velocity, etc., in a medium.
(2) Sound is an auditory sensation evoked by the oscillation described above.

Sound Absorption
Sound absorption is the property possessed by materials and objects, including air, of converting sound energy into heat energy.

Sound Absorption Coefficient
The sound absorption coefficient of a material is the fraction of incident sound energy absorbed or otherwise not reflected by the surface. Unless otherwise specified, a diffuse sound field is assumed.

Sound Pressure
The sound pressure at a point is the total instantaneous pressure at that point, in the presence of a sound wave, minus the static pressure at that point.

Sound Pressure Level
The sound pressure level, in decibels, of a sound is 20 times the logarithm to the base 10 of the ratio of the sound pressure to the reference pressure. The reference pressure shall be explicitly stated and is defined by standards.

Sound Transmission Class (STC)
A rating of the laboratory-derived airborne sound transmission loss data for a structure determined in accordance with the procedures of ASTM E90 and E413. A single number rating of partition airborne sound transmission loss across 16 one-third octave bands between 125 Hz and 4000 Hz as measured in an acoustical laboratory under carefully controlled test conditions. The STC is used during building design phase to select a particular partition/window configuration to obtain desired sound isolation performance.

Sound Transmission Coefficient (τ)
The sound transmission coefficient of a partition is the fraction of incident sound transmitted through it. Unless otherwise specified, transmission of sound energy between two diffuse sound fields is assumed.

1.6.3 Sound & Vibration - Glossary, Abbreviations & References

Sound Transmission Loss
Sound transmission loss (TL) of a material or building partition is a measure of sound isolation ability. Expressed in decibels, it is 10 times the logarithm to the base 10 of the reciprocal of the sound transmission coefficient of the partition. Mathematically, this is represented as:

$$TL = 10\log\frac{1}{\tau}$$

Unless otherwise specified, the sound fields on both sides of the partition are assumed to be diffuse.

Speech Privacy
"Techniques to render speech unintelligible to casual listeners" defined in ANSI T1.523-2001/Glossary, a standard maintained by the U.S. Department of Commerce, National Telecommunications and Information Administration, Information Security Program (INFOSEC). This definition matches earlier ones in ANSI S3.5 (1969) and ASTM E1130 (1997 & 2001). See ASTM E1130 for three defined, measurable levels of speech privacy.

Spectrum
A group of sound levels in frequency bands covering a wide frequency range. Generally, this term is used with some modifier indicating the resolution bandwidth, e.g., octave band spectrum or one-third octave band spectrum.

Stiffness
Stiffness characterizes the ability of a material to resist bending.

Thermal Double Glazing or Insulating Glass
Two monolithic glass panels set into a common frame providing an air space between the two panels that is generally not larger than ¾". Such glazing units are most often hermetically sealed.

Wave
In acoustics, physical disturbance, usually a pressure disturbance, which propagates through a medium, typically air, and building components.

1.6.3 Sound & Vibration – Glossary, Abbreviations & References

PARTIAL LIST OF ABBREVIATIONS

AAP
American Academy of Pediatrics.

AIA
American Institute of Architects.

AHA
American Hospital Association.

ANSI
American National Standards Institute.

ASA
Acoustical Society of America, http://asa.aip.org/

ASHE
American Society of Hospital Engineers, a division of the American Hospital Association.

ASHRAE
American Society of Heating, Refrigeration and Air-Conditioning Engineers.

ASTM
American Society for Testing and Materials.

EPA
U.S. Environmental Protection Agency.

FAA
Federal Aviation Administration.

1.6.3 Sound & Vibration - Glossary, Abbreviations & References

FGI
Facility Guidelines Institute, manager of the Health Guidelines Revision Committee (HGRC).

FHWA
Federal Highway Administration.

GGHC
Green Guide for Health Care, a voluntary rating system for healthcare facilities that is based on the USGBC's LEED HC. The Green Guide for Health Care served as the "pilot phase" for the USGBC's testing of the acoustical criteria in this manual.

HIPAA
Health Insurance Portability and Accountability Act, administered by the Office for Civil Rights of the U.S. Department of Health and Human Services.

HUD
U.S. Department of Housing and Urban Development.

INCE
Institute of Noise Control Engineers, http://www.ince.org.

JASA
Journal of the Acoustical Society of America.

LEED HC
A program designed and administered by the U.S. Green Building Council (USGBC) known as the Leadership in Energy and Environmental Design Health Care Initiative.

NCAC
National Council of Acoustical Consultants, http://www.ncac.org.

OCR
Office for Civil Rights, a division of the U.S. Department of Health and Human Services that is responsible for administering HIPAA.

1.6.3 Sound & Vibration – Glossary, Abbreviations & References

REFERENCES

New references:

- AAP Committee on Fetus and Newborn, Levels of Neonatal Care, Pediatrics 2004;114;1341-1347.

- Cavanaugh, W.J., Tocci, G.C., Wilkes, J.A., Architectural Acoustics Principles and Practice (second edition), Wiley, 2010, ISBN 978-0-470-19052-4

- Florida Building Code. Section 553.73, Fla Stat

- Philbin MK and Evans JB. Standards for the acoustic environment of the newborn ICU. J Perinatol. 2006;26:S27-S30.

- www.SpeechPrivacy.org - website of the American National Standards Institute (ANSI) committee S12-WG44 (Healthcare Acoustics and Speech Privacy) and the Joint ASA/INCE/NCAC Subcommittee on Healthcare Acoustics & Speech Privacy

Printed by Publishers' Graphics LLC